AF327542

ETHYLENE
BASIC CHEMICALS
FEEDSTOCK MATERIAL

ETHYLENE
BASIC CHEMICALS
FEEDSTOCK MATERIAL

Oscar G. Farah
Robert P. Ouellette
Ralph C. Kuehnel
Mario A. Muradaz
Paul N. Cheremisinoff

ANN ARBOR SCIENCE
PUBLISHERS INC
P.O. BOX 1425 • ANN ARBOR. MICH. 48106

FOREWORD

Ethylene ranks first in production among the major organic chemicals, and its consumption is approximately twice that of its next two contenders —propylene and benzene. The ethylene business is capital-intensive and complex, and its stakes are high and continually increasing. Unsettled conditions and problems have made it the subject of concern to the chemical processing industry.

The work presented here originated as a contract study by The MITRE Corporation, sponsored by and prepared for Société Nationale ELF Aquitaine de France. We gratefully acknowledge its support, encouragement and permission to publish this material.

Oscar G. Farah
Robert P. Ouellette
Ralph C. Kuehnel
Mario A. Muradaz
Paul N. Cheremisinoff

Oscar G. Farah is Manager of International Programs, Energy, Resources and the Environment Division at Metrek, a division of The MITRE Corporation. Dr. Farah started his career at MITRE in 1969 and since then has been associated with a variety of projects covering communications, transportation, and energy and the environment. Since 1974, his interests have been concentrated in the area of energy and the environment, and in 1976 he started working in MITRE's international programs. Before joining MITRE he was a research scientist at the Naval Research Laboratories (NRL) and the Carlisle Barton Laboratory of Johns Hopkins University. Prior to that he was a member of the technical staff at Bell Telephone Laboratories and the Western Electric Company. On a number of occasions he has worked as an independent consultant for the Hudson Institute, NRL, Scope, Inc., and The MITRE Corporation. Dr. Farah obtained a BS in Mechanical Engineering from Purdue University and a PhD in Electrical Engineering from Johns Hopkins University. He has published a number of papers on world energy demand and supply, energy transmission and energy conservation.

Robert P. Ouellette is Technical Director of the Environment Division of The MITRE Corporation. Dr. Ouellette has been associated with MITRE in varying capacities since 1969 and has been Associate Technical Director since 1974. Earlier, he was with TRW Systems, Hazelton Labs, Inc. and Massachusetts General Hospital. He graduated from the University of Montreal and received his PhD from the University of Ottawa. A member of the American Statistical Association, Biometrics Society, Atomic Industrial Forum and the NSF Technical Advisory Panel on Hazardous Substances, Dr. Ouellette has published numerous technical papers and books on energy and the environment. He edited the comprehensive survey, *Electrotechnology*, published in 1978 by Ann Arbor Science Publishers, Inc.

Ralph Kuehnel received both an MBA in Quantitative Methods and a BS in Accounting from the State University of New York. At MITRE, Mr. Kuehnel has been responsible for the design and development of the Metrek Model for Financial Evaluation of Boiler Conversion to Coal-Oil Mixtures. Mr. Kuehnel previously worked with public utilities and government agencies in the development of long-range utility plans. These tasks included the design and responsibility for the development of a number of computer models to assist in evaluating economic, environmental and regulatory policies related to utilities. Mr. Kuehnel has also served as financial analyst in the design of a detailed financial performance model for the cable television industry. Under the sponsorship of the Canadian Department of Communications, he directed programming and installation of this model in Ottawa, where it is now being used for policy evaluation, rate regulation and financial analysis of cable systems.

Mario Muradaz is currently working to obtain a degree in Economics from the University of Maryland. At MITRE, Mr. Muradaz has been responsible for research in support of various energy technology projects. His tasks have involved identifying pertinent energy research currently conducted at national laboratories, as well as pertinent Congressional Acts applicable to energy-related activities.

Paul N. Cheremisinoff is Associate Professor of Environmental Engineering at the New Jersey Institute of Technology. He is a consulting engineer and has been a consultant on environmental/energy/resources projects for MITRE Corp. A recognized authority on pollution control, he is author/editor of many publications, including several Ann Arbor Science handbooks on pollution and energy, such as *Pollution Engineering Practice Handbook; Carbon Adsorption; Environmental Impact Data Book; Industrial & Hazardous Wastes Impoundment; Environmental Assessment & Impact Statement Handbook*. Mr. Cheremisinoff is also Engineering Editor of *Water & Sewage Works*, an international journal.

CONTENTS

Chapter .... Page
1. Overview 1

2. Technological Improvements in Ethylene Production 5
   Introduction 5
      Improvements of Conventional Processes 5
      Radical Changes in the Process 6
   Direct Cracking of Crude Oils 6

3. The Ethylene Industry and Its Sources of Supply 13
   Introduction 13
   New Sources of Raw Materials for Petrochemicals 14
   Sources of Raw Materials for Ethylene Production 17

4. Projections and Uses 25
   Introduction 25
   Trends in U.S. Ethylene Production 28
      Yield 31
      Major Trends 32
   The World Ethylene Industry 35

5. Alternative Feedstocks 37
   Introduction 37
   Independent Variables 38
   Definition of Processes Used 40
      Ethylene from Crude Oil 40

Ethylene from Sugar Cane 41
Acetylene from Coal (Arc Process) 43
Acetylene to Vinyl Chloride Monomer 45
Ethylene to Vinyl Chloride Monomer 47

6. Cost Comparisons 49
General Procedure 49
Ethylene from Naphtha 51
Ethylene from Cane Sugar 53
Acetylene from Coal 55
VCM from Acetylene 56
VCM from Ethylene 56
Preparation and Presentation of Results 58
Discussion 67

7. Impact of Technology Improvement 71
Introduction 71
Ethylene from Crude Oil 71
Ethylene from Biomass 72
Ethylene/Acetylene from Coal 72

8. Naphtha Feedstock Price Projection 75
Introduction 75
Theory of Refinery Cost Allocation 75
Product Price Forecasts 81

9. The MITRE Full Life-Cycle Cost Model 85

10. Historic Cost/Price Trends 93

References 97

Bibliography 99

Index 101

LIST OF FIGURES

Figure 2-1. Diagram of experimental Advanced Cracking Reactor. . 8
Figure 2-2. Formation of cracked gases (low ethylene:acetylene). .. 9
Figure 2-3. Formation of cracked gases (high ethylene:acetylene). . 9
Figure 4-1. End uses of ethylene. 26
Figure 4-2. End use of acetylene. 27
Figure 4-3. Future ethylene feedstock sources. 30
Figure 4-4. Feedstocks for ethylene in the United States. 31
Figure 4-5. U.S. ethylene industry, 1976. 33
Figure 4-6. World ethylene feedstock and capacity. 34
Figure 5-1. Steps utilized in arriving at the cost of ethylene from
the three feedstocks. 38
Figure 5-2. Ethanol from sugar cane. 42
Figure 5-3. Diagram of ethylene from ethanol plant. 43
Figure 5-4. Schematic of coal conversion arc process and
coal reactor. 44
Figure 5-5. Diagram of arc-coal plant. 46
Figure 5-6. The balanced oxychlorination process for the
production of VCM from ethylene. 48
Figure 6-1. Illustrative chemical process cost flow. 50
Figure 6-2. Material balance for the manufacture of VCM
from acetylene. 57
Figure 6-3. Material balance for the production of VCM
from acetylene. 58
Figure 6-4. Cost of VCM as a function of the cost of acetylene. ... 65
Figure 6-5. Cost of VCM as a function of the cost of ethylene. 66
Figure 6-6. Cost of ethylene produced by various processes. 68
Figure 8-1. Topping refinery schematic. 76
Figure 8-2. Hydroskimming refinery schematic. 77
Figure 8-3. Fuels refinery schematic. 77
Figure 8-4. Conversion refinery schematic. 78
Figure 8-5. Refining and petrochemical process facilities total
erected cost index. 82
Figure 9-1. Comparison of approaches to levelized cost. 87

Figure 9-2. Life cycle costs. 88
Figure 9-3. Life cycle cost recovery. 89
Figure 9-4. Life cycle costs. 90
Figure 9-5. Life cycle cost recovery. 91

LIST OF TABLES

Table 2-1. Comparison of Yields of ACR and Conventional Naphtha Cracker 10
Table 4-1. Projected U.S. Gas Liquids Supply 29
Table 4-2. Heavier Feedstocks Predominate for New U.S. Ethylene Capacity 30
Table 4-3. Olefin Plant Yields 32
Table 4-4. Relative Olefin Plant Yields 33
Table 4-5. European Ethylene Capacity Growth 35
Table 5-1. Annual Olefin Plant Operating Costs 39
Table 5-2. Olefin Plant Yields Naphtha Feedstock 40
Table 6-1. Financial Parameters 52
Table 6-2. Independent Variable Costs 52
Table 6-3. Ethylene from Naphtha Fact Sheet 53
Table 6-4. Allocated Equilibrium Product Values as Dependent on Crude Price 54
Table 6-5. Ethylene from Ethanol Fact Sheet 54
Table 6-6. Acetylene from Coal Fact Sheet 56
Table 6-7. VCM from Acetylene Fact Sheet 57
Table 6-8. VCM from Ethylene Fact Sheet 58
Table 6-9. Calculated Product Costs 59
Table 6-10. Cost of Ethylene Produced from Naphtha and from Ethanol 60
Table 6-11. Variations in the Cost of Acetylene as a Function of Variations in the Cost of Coal and Electricity 61
Table 6-12. Percent Cost Distribution in the Production of Ethylene from Naphtha and from Ethanol 62
Table 6-13. Percent Cost Distribution in the Production of Acetylene from Coal 63
Table 6-14. Percent Cost Distribution in the Production of VCM from Acetylene and from Ethylene 64

Table 6-15. Cost of VCM 65
Table 6-16. Cost of Ethylene 66
Table 8-1.   Olefin Plant Yields 76
Table 8-2.   Illustrative Cash Flow Summary for Conversion
             Refinery 79
Table 8-3.   Equilibrium Price Relationships 81
Table 8-4.   Arabian Light Crude Assay 82
Table 8-5.   Refinery Yields from Arabian Light Crude Oil 83
Table 8-6.   Allocated Equilibrium Product Values as Dependent
             on Crude Price 84
Table 10-1. Price History of Selected Compounds Used in the
             Petrochemical Industry 94

CHAPTER 1

OVERVIEW

INTRODUCTION

This book examines the future of the U.S. ethylene industry by considering its alternate sources of supply, the expected technological improvements in its production processes, its demand growth, and its available and under-construction production capacity. This study was based on a survey of the literature and on industry contacts.

Using information on energy resources, the background necessary to permit a logical forecast of the future of the ethylene industry is presented. Two alternative scenarios are given:

1. continued oil glut and expansion of the availabiity of natural gas; and
2. business as usual, but with spot shortages and curtailments.

In either situation the ethylene industry should have no problem surviving, and its principal feedstocks will continue to be crude oil and natural gas. Principal difficulties (up to 1985) will be due to present overproduction capacity, drop in demand growth from 12% to 6.5–7% and the coming onstream of ethylene plants in OPEC countries, Canada and South America. With either scenario, the quantity of coal- and biomass-derived petrochemicals will be negligible and could provide no more than 10–15% of production by the year 2000. However, the technologies for utilizing coal and biomass cannot be neglected as these feedstocks do provide "an insurance" against possible long-term shortages. Biomass technology appears to offer an advantage to the less-developed countries that are short on capital and require labor-intensive industry.

An important threat in the background is the possible disruption of production in the oil-producing countries due to political causes. Internal

1

unrest in these countries could cause immeasurable damage to world economies if allowed to surface. Short disruptions could be handled with minimum difficulty, but for long disruptions of supply, the total world economy would suffer.

Alternate sources of supply are being developed: the heavy oils and tar sands of Canada and Venezuela; the oil resources of Third World countries; the immense reserves of natural gas in geopressured reservoirs in Louisiana and Texas; and the oil reserves of Mexico and China. But under present, nonwar conditions, development of these resources will be slow because survival is not at stake. With the passage of the modified Carter energy bill, new economic conditions now exist in the United States that are conducive to increases exploration, and future availability of feedstocks appears somewhat assured.

The principal difference between the two hypothesized scenarios is the cost of feedstock. In an oil glut and expansion situation, the cost should decrease in real terms; with business as usual but curtailment, the cost would increase. The cost increase should be modest, unlike that of 1973–1974, and will effectively be healthy on the supply situation. Some of the more expensive oil and gas resources would become economical, e.g., tertiary recovery, the oil shales, tar sands, etc.

To help ensure the survival of the ethylene industry it should continue (1) to be prepared to accept a range of feedstocks; (2) to develop preconditioning technology to hydrogenate the heavier fractions to increase ethylene yield; (3) to develop new processes to crack crude directly that can accept a range of fractions; and (4) to maintain a healthy R&D program in alternate raw materials (coal and biomass).

OVERVIEW

The U.S. Chemical Processing Industry market conditions, and specifically those for ethylene prior to the fourfold increase in oil prices, were as follows:

1. Ethylene had been and still is so basic to the petrochemical industry that its rate of growth had been correlated with the growth rate of the Gross National Product (GNP). In the 1960s, ethylene growth rate was 3–3.5 times that of the GNP. Between 1960 and 1974 the growth rate of ethylene demand averaged 12%.

2. The demand for ethylene appeared to be unlimited as most natural products such as wood, paper and metals were being replaced by plastics, which were ethylene derivatives.

3. Natural gas, which had been the principal feedstock in the production of ethylene, was becoming scarce, and planners were having serious doubts about the reliability of supply; the relatively low price of natural gas, which

> was set and regulated by the U.S. government, made it uneconomical for the natural gas industry to look for new sources of natural gas in the United States.
>
> 4. The technology for production of ethylene from crude oil was available and was widely used in Europe. The conventional process consisted of cracking the crude to produce naphtha and using the naphtha as a feedstock for producing ethylene. Naphtha cracking produced a range of additional coproducts that had to be disposed of. Although the crude oil route was more expensive than the natural gas route, the coproducts were valuable to the refiner.
>
> 5. The technology for producing ethylene and the economies of scale had reached a point where small ethylene plants were no longer economical and the size of an economical world-size ethylene plant was now on the order of one billion lb/yr.

Faced with these facts and taking into account the long planning and construction lead times, the CPI went ahead bullishly with major expansion plans. (It is the opinion of some CPI experts that the CPI industry did foresee the slump in ethylene demand; however, because of the expected increases in capital costs to construct ethylene plants, the oil companies deliberately opted to increase their capacity and overbuild before construction costs skyrocketed.) The scarcity of natural gas forced the shift to crude oil, and chemical companies, which owned the majority of the natural gas-based ethylene plants, found the ownership of these plants unattractive because they had to market a range of coproducts. The oil industry, which already had markets for these products, took over the production of ethylene. Union Carbide, in cooperation with Kureha Chemical Company and Chiyoda Kako Kensetsu of Japan, chose to develop a process of cracking crude oil directly to produce ethylene using a new Advanced Cracking Reactor (ACR).

Then came the unexpected fourfold oil price increase and general slowdown of the economy in 1975. Today, with prices somewhat stabilized, the demand for ethylene is rising, but seems to have stabilized at slightly more than twice the GNP growth rate.

If one were to ask CPI experts their opinion on the future of the ethylene business, one would be surprised by the diversity of opinions received. For example, the marketer would paint a gloomy picture of this future based on the available overcapacity for ethylene production; the marketing researcher or planner would argue that based on present growth rates, at least 20 new world-size plants would be required by 1985; the supply specialist would insist that ethane/propane are no longer available and it is necessary to switch to oil; while the explorer would counter that twice the U.S. coal reserves in gas is available in the geopressured reservoirs of Louisana and Texas; also, the researcher would maintain that shale or lignite would provide excellent feedstock for ethylene production when using a transfer line or riser for cracking; and finally, the refining and production specialist would stress that the naphtha route is unattractive because as more severe reforming is used to increase the yield of gasoline, naphtha yields would decrease. Finally, the coal and biomass advocates would insist that future petrochemicals should come solely from their respective raw material.

CHAPTER 2

TECHNOLOGICAL IMPROVEMENTS IN ETHYLENE PRODUCTION

INTRODUCTION

The CPI is continuously conducting research and development to find more efficient processes. Because of the large capital investments required to construct world-scale plants and the narrow profit margins, a new plant must be of the most efficient, highest yield design. In ethylene production, improvements can be classified into two general areas.

Improvements of Conventional Processes

These improvements are made gradually, and the resulting yields inch up slowly. The improvements are generally minor and time consuming. In ethylene production, ethylene is conventionally obtained by pyrolysis of hydrocarbons. High selectivity for the desired olefins and diolefins and minimum coke production can be achieved by operating at high temperatures, short residence times and low hydrocarbon partial pressures. Hence, the components that have the most effect on ethylene yield are the reactor and the quenching unit.

Pyrolysis takes place in the reactor, which is essentially a heated vessel. The upper temperature limit is set by the material from which the vessel is made. Uniform temperature in the vessel is essential to ensure isotropic pyrolysis and minimum coke deposition. For this reason, the tubular reactor has emerged as the most effective design and can be used at elevated temperatures to achieve high-severity pyrolysis with naphtha feedstocks. Pyrolysis temperatures up to 1000° C can be achieved with presently

5

available materials. Under these conditions, up to 34.4% of ethylene production can be obtained from a naphtha feedstock.

The quenching unit is instrumental in controlling residence time. This unit must provide rapid cooling of the reactor effluents to a temperature where further reactions do not occur. In older plants, the effluent was quenched to an intermediate temperature with water and cooled further in transfer line heat exchangers. More recently, with improved heat exchanger design and efficiency, heat recovery has been maximized by eliminating the water quench and quenching in the transfer-line exchangers. Improvements in the design and efficiency of the heat exchangers are of importance in the future.

Radical Changes in the Process

In this area, new processes are designed that are radically different. In ethylene production, the most noteworthy investigations have been the study of direct production of ethylene from crude and catalytic cracking to increase the yield of ethylene. Of these, direct ethylene production appears to be advantageous under certain conditions and will be discussed here. Catalytic cracking, on the other hand, does not appear to offer sufficient improvements to warrant its utilization. The consensus of the limited experimentation reported in the literature indicates that catalytic cracking would offer small improvements in ethylene yield, which would not be viable from an economic standpoint.

DIRECT CRACKING OF CRUDE OILS

Direct cracking of crude oils has been achieved in the Advanced Cracking Reactor (ACR). The ACR can crack crude oil and produce ethylene directly without the conventional intermediate step of naphtha production. Unlike the naphtha cracker, the ACR will yield 70% chemicals based on crude, while the conventional technique yields only 50%. As such, it is an ideal process for the chemical industry, which is not interested in producing a range of heavy coproducts that must be disposed of in an unfamiliar market.

While the ACR was originally intended to crack whole crude oil, it is now being modified to handle deasphalted crude and other refinery products. Deasphalting is carried out by a distillation step, similar to conventional refining. This step yields vacuum gas oil, which is the charge to the ACR, and asphalt, which is burned as a fuel. Cracking is performed at high temperature and pressure and low residence time. A diagram of an experimental ACR is shown in Figure 2-1. The deasphalted crude oil is injected into somewhat less than two times its weight of high-temperature gases and superheated steam at high pressure at about 2000° C. For the experimental unit, the gases are generated by combustion of excess fuel with pure oxygen generated by a rocket booster.

The gases, steam and feedstock droplets are accelerated through a venturi nozzle reaction chamber where adiabatic cracking occurs. The reaction products are then quenched rapidly after a residence time of about 20 msec by a new type of heat exchanger, which is a key factor in the ACR's feasibility.

This heat exchanger is called the "Ozaki quench cooler" after Kiyoji Ozaki, who helped develop the technique. Prior to the development of this cooler, direct cracking was always plagued with coking problems in the cooling section. Dow Chemical, in its attempt to crack crude oil in the 1940s, utilized a twin reactor system that necessitated the burning off of the carbon deposited on the walls of one reactor while the other was being used.

In the Ozaki system, a quenching oil obtained from the gasoline fractionator is jetted into the high-temperature reaction product stream to precool the gas. This mixed phase (steam, hydrocarbon gases and quench oil) is fed into a tubular heat exchanger, where the heat is recovered to make high-pressure steam. The steam and quenching oil flow along the inside wetted surface of the exchanger tubes, preventing the high-temperature gases from coking the walls and improving the coefficient of heat transfer. Also, since the gas is mixed with the oil, the film formed on the tube surface is thinner, resulting in an improved heat transfer coefficient.

During an experimental run, the system was run for 960 hours with an Arabian light crude with very little coking. The overall coefficient of heat transfer during this run remained in the range of 350–400 kcal/m^2–° C. The liquid film also protected the heat exchanger from corrosion by hydrogen sulfide. The high-temperature coefficient allows the production of high-pressure steam of 40–100 kg/cm^2 and results in a decrease in the size of the heat exchanger.

The quenching oil has high stability and can accept pitch concentrations of up to 80%. By controlling the reaction conditions, ACR operation can be tailored to produce a variety of product mixes, as shown in Figures 2-2 and 2-3.

In Figure 2-2, the reaction conditions are programmed for a low ethylene:acetylene production ratio, while in Figure 2-3, the reaction conditions are set for high ethylene:acetylene production ratios. For high acetylene production, the downstream gas-separation train requires major modifications to handle the explosive acetylene concentrations.

The quenched reactor products are fed into a gasoline fractionator that

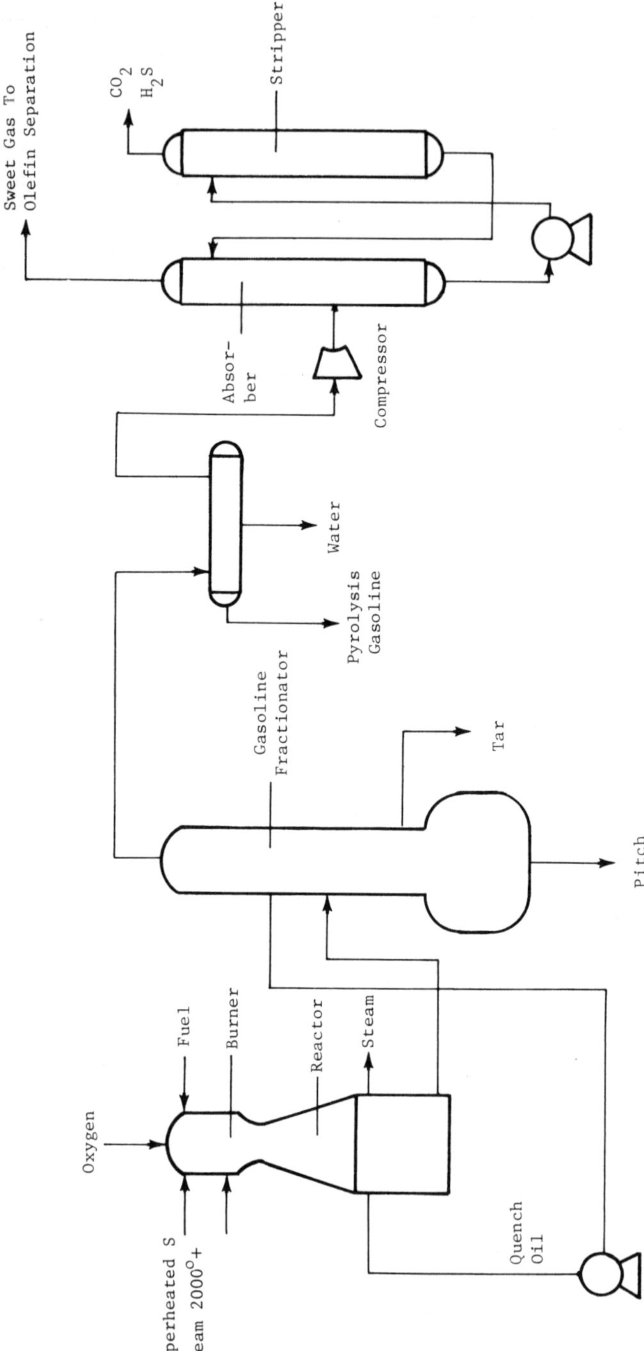

Figure 2-1. Diagram of experimental Advanced Cracking Reactor (Gomi and Wilkinson, 1974; *Chem. Eng. News*, March 27, 1978).

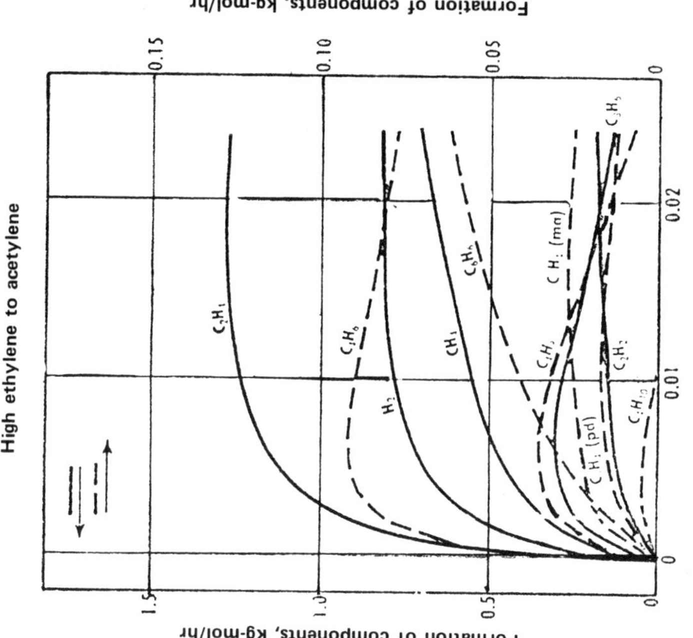

Figure 2-3. Formation of cracked gases (high ethylene:acetylene) (Gomi and Wilkinson, 1974).

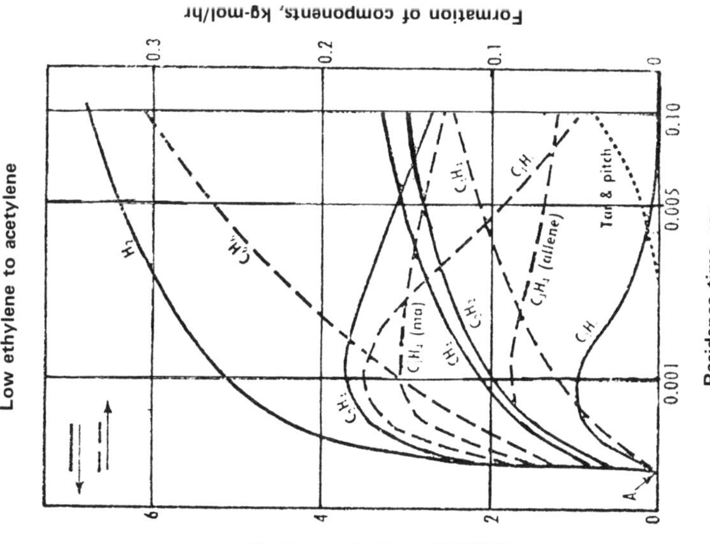

Figure 2-2. Formation of cracked gases (low ethylene: acetylene) (Gomi and Wilkinson, 1974).

separates the heavy tars, lighter liquid fractions and gaseous products. The gaseous products are compressed and scrubbed free of carbon dioxide and hydrogen sulfide in a removal system designed to cope with the relatively high concentrations of polymer formers in the cracked gas. This sweet gas is then refined in a gas-separation train (not shown in Figure 2-1) adapted to the cracked gas composition. This train is adapted to include an acetylene purification system and modifications to limit concentrations of acetylenic compounds and polymer formers. The acetylene produced can be used directly as a feedstock to a VCM plant or may be hydrogenated to produce olefins.

Table 2-1 compared the yields of a typical commercial ACR to that of a conventional naphtha cracker. It is apparent from this table that the ACR offers great advantages for a chemical company whose principal interest is the production of hydrocarbon chemicals. The crude oil requirements are cut by a factor of more than five, and for the same amount of C_2 production, the crude oil required is more than six times less. This leads to greater stability of return on investments. The price of ethylene produced in this fashion is less sensitive to changes in raw material and by-product costs. In terms of ethylene price, although the ACR capital costs are expected to be larger than those for a naphtha cracker, the improved efficiency is expected to result in a savings of 3–4 ¢/lb of ethylene. (This is principally because of the high demand for naphtha in gasoline production.)

However, no commercial ACR are in operation as of this writing. Prototypes have been built to demonstrate its feasibility, but it is not certain that the model would be successful. The scaleup problem is in

Table 2-1. Comparison of Yields of ACR and Conventional Naphtha Cracker
(Gomi and Wilkinson, 1974)

| | ACR | | (Typical) Naphtha Cracking | | |
| | | | | Naphtha | Crude Oil |
	10^6 lb	%	10^6 lb	%	%
Raw Material					
Crude Oil	3050	100	(16,000)	423	100
Naphtha	–	–	3780	100	23.6
Products					
Ethylene	1000	32.8	1000	26.4	6.2
Acetylene	200	6.5	15	0.4	–
Propylene	135	4.4	590	15.6	3.7
Crude Butadiene	90	2.9	359	9.4	2.2
Crude BTX	435	14.2	990	26.2	6.2
Pitch	583	19.1	–	0.0	0.0
Fuel	600	19.7	840	22.2	5.2

maintaining chemical similarity in the larger reactors. Unlike mechanical, thermal and dynamic similarity, chemical similarity is not well defined. Computer simulations of the reaction zone indicate that the vapor-phase cracking reactions are much faster than the vaporization, which is substantially complete by the time the throat of the venturi is reached. The scaling problem is complicated by a mixed flow regime in the portion of the reactor upstream of the throat. The pyrolysis rate depends on the temperature profile in the reactor, while the rate of bulk flow depends on the flow pattern itself.

In view of possible difficulties, Union Carbide elected to defer the construction of a world-size plant and chose instead to construct a $15 million intermediate-size prototype in Seadrift, Texas, primarily to demonstrate long-term equipment operability. The unit was scheduled for completion in 1979 and, if successful, will lead to the construction of a world-scale ethylene plant in the mid-1980s.

Dow Chemical is also experimenting with crude cracking in Europe. The process is similar to Union Carbide's in that it utilizes high temperature and short residence time. It involves burning part of the crude feed with oxygen, and the stream is then rapidly quenched. The acetylene:ethylene ratio produced is 1:15, and the ethylene yield is 26–29% of the crude feed, but with a higher ethylene:propylene ration (3–4:1). Experimental runs in a small prototype of long duration have been performed without any coking difficulties.

Assuming that Union Carbide and Dow develop processes that can produce ethylene directly from crude oil at a cost lower than by naphtha cracking, all future supplies of ethylene will be produced in this manner. Obviously, direct cracking routes offer large advantages for the chemical company:

1. They can purchase crude oil directly and eliminate their dependence on the refiners. Also, because the process is not limited to the use of crude oil but can use almost any feedstock, it is a decisive economic advantage. For example, heating oil could be used in the summer months if crude supplies became tight because of the higher gasoline demand. They could conceivably have their own agreements with the oil producers to import crude.

2. They can maintain price stability for ethylene because of the higher yield and resulting decrease of the by-products.

3. They can develop a petrochemistry based on low-priced acetylene.

4. They can utilize the tar and fuel produced to fuel other processes. In the case of Union Carbide, it can utilize the coke produced, which it needs for other processes and credit itself with a high premium price for this product.

Of course the advantages listed above are specific to large chemical companies with diversified products. These are not necessarily advantages for the small chemical companies. To the refiners who use only 5% of the crude barrel for petrochemicals, the direct cracking route is certainly not in their interest. The high pitch yield and the very high sulfur concentration in the fuel make the use of these coproducts (in excess of 35%) problematic. The sulfur content may be controlled by using the more expensive sweeter crudes, such as the Libyan and Indonesian oils. But these can conceivably be run directly into the conventional naphtha crackers.

Hence, except for large chemical companies, the direct cracking route is not the route of choice. Rather, the preferred route is to improve the conversion efficiencies of their present processes to produce the full range of products and decrease the proportion of heavy bottoms. In summary, direct cracking technology is a selective tool for large chemical manufacturers who want to stabilize their source of supply and eliminate their dependency on the refiners.

CHAPTER 3

THE ETHYLENE INDUSTRY AND ITS SOURCES OF SUPPLY

INTRODUCTION

Recent estimates of the growth of the U.S. ethylene industry have been placed at between 6.5% and 7% per year, as compared to an average of 12% for the 1960–1974 period. Although higher feedstock prices and the slowdown in the economy can be blamed for this decrease, the demand for ethylene and its derivatives is reaching saturation. The original high growth rates were due to the substitution of natural products such as wood, fibers, rubber, paper and metals by products derived from ethylene. It appears that the initial substitutions with their associated large growth rates have been made and although further substitutions are still possible (e.g., in the auto industry), their impact will not be as great as it was earlier. The decreased population growth rates will also contribute to the slower growth rate of ethylene demand. Higher oil prices and the slowdown in the economy did have an indirect effect on ethylene demand as they resulted in the postponement of planned expansions of industries that would have ultimately used ethylene and its derivatives.

On the other hand, the ethylene industry, which has been geared for high growth rates, had built some $2 billion worth of new facilities for ethylene, propylene and butadiene production; also, another $3 –4 billion worth of new olefin plants are now under construction. This resulted in the availability of large surplus capacity for ethylene production. Decreased demand for ethylene-derived products (plastics) has resulted in delays in the construction of these plants, which left a number of ethylene production plants with no outlet for their products.

This has meant a softening of ethylene prices in the U.S. In 1977, U.S. ethylene demand was 80% of the available 31 billion pound capacity; in 1978, this percentage was 78% of the 34 billion pound capacity then available. As a result, the planned construction of additional ethylene capacity is being postponed. However, projects under construction at present are expected to add 8 billion pounds of capacity by 1982. With the present growth rate, it will not be until 1985 that the available capacity would become necessary.

To further complicate matters, the possible introduction of new processes by Union Carbide and Dow Chemical that crack crude oil directly and yield lower priced ethylene should result in further softening of the ethylene market. Additionally, it is expected that by 1985 new production facilities now under construction in Canada, Mexico, South America and the OPEC countries will come onstream and decrease the demand for exports from the U.S., and possibly provide competition to U.S. production facilities. Also, Europeans are constructing ethylene plants that would use natural gas liquids from Algeria and the North Sea.

In the U.S., the oil companies will carry the brunt of excess U.S. ethylene capacity. As a result of natural gas shortages, most new ethylene production plants use naphtha cracking and are owned by the oil companies. Most ethylene facilities that utilize natural gas are owned by the chemical companies, which have internal uses for their production and will only purchase ethylene from outside sources when their demand exceeds their production.

Although the CPI normally looks for a after-tax profit of 10–15%, it will have to be satisfied with smaller profits to keep its production facilities in operation. Older plants would make a small profit by running as low as 65–75% of capacity; however, new plants, with their higher initial capital costs, must run at 70–75% to break even. It is expected that the smaller, uneconomical older facilities will be shut down to allow the newer larger facilities to operate.

In conclusion, it appears that the U.S. ethylene market is presently saturated with production facilities that utilize fossil fuels for feedstocks and this situation will probably continue until at least 1985.

NEW SOURCES OF RAW MATERIALS FOR PETROCHEMICALS

Two potential raw materials for the chemical industry are coal and biomass, which are the principal sources of energy utilized by man prior to the discovery of petroleum. The U.S. has enormous coal reserves; the known recoverable portion of these reserves is 250 billion tons, and the ultimate reserves could be twice that. Coal can be used directly to produce energy; however, for the chemical industry, an intermediate product is generally needed. Conversion of coal to an intermediate, whether it be a gas or a liquid, requires heavy capital expenditures. At present, none of these intermediates can compete with the world oil price, and it is postulated that as the price of oil continues to climb and shortages develop, coal-derived feedstocks would become competitive. Historical data on the price relationship between oil and coal indicate, however, that the price of coal increases with that of oil; and it always appears that the price of coal-derived feedstocks is higher than the price of oil. The U.S. has active programs in coal utilization ranging from direct combustion to the production of methanol and acetylene for use in transportation and the chemical industry, respectively. South Africa, which has no oil reserves, is very active in the development of gasification and liquefaction processes.

Conceptually, everything that can be manufactured from petroleum can also be manufactured from coal by adding hydrogen or water. As a matter of fact, synthesis gas, which is a 1:1 mixture of carbon monoxide and hydrogen, is produced in the United States at an estimated rate of 70 billion lb/yr and can be used to replace methane and naphtha in many applications. It is conceivable then that coal could be used to displace petroleum.

Biomass is also a hydrocarbon and could conceivably be used to replace petroleum. Its principal advantage is that it is a renewable resource. Worldwide, it is estimated that the net productivity of forests exceeds in heating value the annual consumption of fossil fuels. The major difficulty with biomass is that much labor and energy must be expended to collect the biomass and transport it to processing centers.

Again, it would be most inconvenient to use biomass directly in the chemical industry, so an intermediate product is required, the most favored of which is ethanol. Biomass is fermented, and the resulting alcohol is then distilled and used as a feedstock. Brazil has a very active program in this area and intends to produce ethanol from sugar cane, cassava and other similar plants to displace 20% of its gasoline consumption. Also, new vehicles that utilize 100% ethanol are being developed. For the chemical industry, a number of designs for ethylene from ethanol plants are now available, and Brazil is constructing a 60,000-metric ton plant.

In other programs, plants that grow in arid areas that contain hydrocarbons capable of serving as a petrochemical feedstock are being studied. Some of these plants include guayule, jojoba, candelilla, milkweed, gopherweed, dogbane and some sunflower species. It is estimated that an acre of gopherweed could produce 10 barrels of oil equivalent.

The cost of ethylene from ethanol is presently at least three times larger than the cost of ethylene from oil, with 85% attributable to the feedstock cost (ethanol). A Columbia University professor, Harry Gregor, has developed a new plastic membrane that could be used to purify and predistill the products of fermentation, resulting in a considerable reduction (about a factor of three) in ethanol's production cost. If this is true, ethylene from ethanol could become a competitive process.

Without this development, biomass probably will only be economically viable if it is already available and must otherwise be disposed of, i.e., where residues are available, such as those from municipal waste and paper production. Here, the biomass material is available at little or no cost; the principal costs are those for capital and operation and maintenance. The Gulf Chemical Company has recently announced the development of an ethanol production process that uses municipal solid waste and agricultural and industrial wastes as feedstocks. It is claimed that ethylene produced in this way could "beat the price of ethylene" produced from petroleum. The process uses new enzymes developed by Gulf that convert cellulose to glucose in hours, instead of the 4–10 days previously needed. The most cost-effective plant size is estimated to be 25 million gal/yr. This could yield less than 100 million lb/yr of ethylene. It would require more than 10 plants to equal the output of one world-size ethylene plant. Depending on the size of the city and the amount of municipal and other waste available, these plants may have to be located in different cities, which could create transportation problems. This indicates that biomass could conceivably pose a threat to oil as a feedstock for ethylene production. This will be explored further later on in this section.

Other emerging sources of raw materials for the petrochemical industry are the bitumens and heavy oils from the tar sands and oil from shale. There is no special merit in the heavy oils becoming a chemical feedstock. Heavy hydrogenation would be required to permit their use. In this case, however, they would become similar to conventional crude oils and would use the same equipment utilized for conversion of crude oil.

Another important, but often neglected, source of raw materials for the chemical industry is natural gas from nonconventional sources. In particular, the natural gas dissolved in the geopressured reservoirs of Louisiana and Texas is estimated to be sufficient to provide all of the U.S. energy needs for at least 200 years. These aquifers are at depths ranging from 8000–26,000 feet and contain brines under very high pressure with large amounts of dissolved natural gas. The extraction of this gas consists of drilling wells into the aquifers processing the brine by passing it through a low-pressure vessel where the gas would be released and disposing of the brine by reinjecting it into the ground to prevent possible settling. It is estimated that this gas would cost more than $4/million Btu, which is twice the U.S. gas price (at the time of writing) and a little more than the price of imported liquefied natural gas (LNG).

Natural gas as a source of petrochemicals and energy is very desirable as it can be utilized directly with little environmental impact. The U.S. is already dotted with natural gas distribution systems, and it would be very simple to add this gas, if extracted, into the present distribution network. Until now, natural gas from geopressured reservoirs has been neglected because of the artifically low price of interstate gas in the U.S. However, as a result of President Carter's modified energy bill, which permitted increases in the price of natural gas, the U.S. Department of Energy (DOE) has initiated studies to explore the possibility of recovering this gas. If the technology for extracting this gas is developed and price reductions result, this gas could pose a serious threat to facilities that produce ethylene from crude oil.

Another source of nonconventional natural gas is from nonbiological origin. It has been postulated that the earth contains large reservoirs of natural gas that were trapped into the interior of the earth during its period of formation. This gas was, and still is, seeping out through the earth. It collects and becomes trapped under impervious rock formations and can be found by drilling almost anywhere where such formations exist. Some evidence is available that could be interpreted to substantiate the existence of this gas. Whether this gas exists is not clear at present, and estimates of the available quantities and cost to extract are not reliable. If it proves to exist in large quantities and can be recovered at competitive prices, it would change present techniques of exploration and have a dramatic impact on the world energy situation generaly and on the petrochemical industry specifically.

SOURCES OF RAW MATERIALS FOR
ETHYLENE PRODUCTION

Perhaps a brief summary of the discussion to this point would be appropriate here:

1. The growth in demand for ethylene appears to have dropped to 6.5–7%.

2. Present U.S. ethylene production capacity is about 34 billion lb/yr, and the demand is 26.5 billion lb/yr. All U.S. ethylene production capacity uses fossil fuels for feedstock.

3. U.S. ethylene capacity will automatically increase to 42 billion pounds by 1982. At that time, ethylene demand will be around 32.4 billion pounds, or 77% of capacity (7% growth rate). By 1985, the demand will grow to 39.8 billion pounds or 95% of capacity, assuming no new plants are constructed.

4. New processes for direct cracking of crude oil will become commercial by 1985 and should add at least 2 billion pounds to capacity (assuming

the construction of only two world-size plants). This means that by 1985, with a 7% growth rate in demand for ethylene, the capacity factor would be about 90%. The new process should cut the cost of ethylene by 3–4¢/lb.

5. Investments in ethylene production facilities are very large and escalating. The cost of a world-size plant could exceed $500 million.

6. New production facilities in Canada, Mexico, South America and the OPEC countries will come onstream in 1985. The output of these facilities will decrease the demand for U.S. exports and may provide competition for U.S. products in the U.S. market.

7. Although the technology for utilizing coal as a feedstock exists, the investments required are very large, and only token productions of ethylene would be made with government support.

8. Developments in biomass technology could allow the construction of ethylene plants based on biomass; however, the production quantities by 1995 would be insignificant.

9. Heavily hydrogenated oil from tar sands and oil shale could provide a substitute to conventional crude oil.

10. Immense reserves of natural gas exist at relatively high cost. Other nonconventional natural gas sources could become available.

The following facts should be added to help complete the picture:

11. The petrochemical industry in the developed countries utilizes a very small portion of the country's oil requirements, and in the U.S., in case of shortages, the petrochemical industry has priority over other industries for oil supplies.

12. Since the large increase in oil prices in 1973, the real price of oil has actually been decreasing and is now about 20% less than it was (in 1973 dollars).

13. The world is now glutted with oil. Recent pronouncements in the U.S. indicate that with the new increases in natural gas prices, a great deal of gas will become available.

14. Mexico has emerged as a very important oil nation. Recent estimates place its reserves at 300 billion barrels, twice the recoverable reserves of Saudi Arabia.

15. China is emerging as an oil nation, and U.S. and other oil companies are actively pursuing agreements with China for exploration.

16. The World Bank has announced that it will make $3 billion available through 1981 to help Third World countries explore for oil and gas with the intent of their becoming self-sufficient.

17. At the time of writing, internal problems in Iran are causing concern over their output of oil. Normally, Iran produces 6 million bbl/day. However, since the 1974 OPEC embargo, the U.S. has initiated a strategic oil storage program. Although this storage program is not progressing as planned, the U.S. can now handle short disruptions in oil production.

18. When hydrocarbons are burned, carbon dioxide (CO_2) is produced. This CO_2 is released into the atmosphere. It is predicted that if coal were to be utilized, the CO_2 content of the atmosphere could reach a level that would turn the earth into a "greenhouse" and could have dramatic effects on world climate. When coal is utilized in place of natural gas, for example, twice as much CO_2 is produced. Although problems are not expected until at least the year 2030, there is a growing movement in the U.S. to curtail the use of coal.

19. Some countries may elect to develop their local resources to substitute for imports. The driving force in this instance may not be economic considerations, but rather national security, to decrease dependence on imports; socioeconomic forces, to provide work for idle labor and redistribute income; or simply for reasons of national pride in developing an industrial base.

Based on these facts, we will now attempt to explore the future of the ethylene industry, first by defining scenarios for fossil fuel availability. At present, two scenarios are plausible:

    I. Continued oil glut and expansion of the availability of natural gas.
   II. Business as usual but with possible spot shortages and curtailments

Other scenarios could be hypothesized. One could be the complete depletion of all oil and gas or a large dramatic step function increase in the price of feedstocks, which effectively results in making feedstocks too expensive and forces a reversion in the trends to substitute natural materials with petroleum-derived materials. Another scenario would hypothesize the discovery of huge low-priced reserves that would bring the price of oil down to pre-1973 prices. In a real world situation these scenarios are highly improbable.

Scenario I

Assuming that natural gas will be available at a price competitive with crude oil prices, the following arguments can be made concerning the evolution of the ethylene industry:

1. The price of oil will remain constant and decrease in absolute terms.

2. Increased availability of natural gas should make it an attractive feedstock for ethylene production. Ethylene from natural gas requires smaller capital investments, and new expansion in ethylene capacity will certainly be by the construction of natural gas-based plants. This would lessen the demand for naphtha and make it more available for gasoline production. However, present ethylene capacity exceeds the demand, so new plants based on natural gas will not be required until 1985.

3. Increased availability of natural gas and the existence of an oil glut will eliminate the urgency for the development of direct crude cracking. Also, decreased demand for naphtha would decrease the economic advantage of the direct cracking route. Since the direct cracking route is especially attractive to the chemical industry, the increased availability of natural gas would permit these companies to revert to their previous modus operandi—ownership and operation of ethylene plants.

4. Decreased dependence on naphtha cracking would create shortages in the important coproducts, such as propylene, butadiene and BTX, which had growth rates comparable to that of ethylene. Recently, however, the growth of these products has been lagging behind the growth of ethylene. This would necessitate the importation of these products from Europe and Canada. However, since Europe is gearing up for natural gas from Algeria and the North Sea, it may not be able to supply these products. Canada, on the other hand, is constructing excess naphtha cracking capacity, with the intent of exporting their surplus to the U.S. However, in 1978, unused capacity for manufacture of these products was more than 23% for propylene, 16% for butadiene, 53% for benzene and 22% for xylenes. This unused capacity was expected to increase.

5. Coal-derived feedstocks will remain in the background and could be made available in the U.S. with heavy government subsidies.

6. Recent advances in biomass technology, if substantiated, will make ethylene from ethanol a viable alternative. However, the relatively small world-size plants would only provide sufficient output to partially satisfy regional needs. The important factor in biomass technology is its dependence on renewable resources, as it is important for the developed countries to master the technology and have it available. Ethylene from biomass, even without improvements in technology, is important to the developing countries: it saves them hard currency and employs their otherwise idle labor force. This is the principal goal of Brazil's PROALCOOL program.

Scenario II

This scenario means that the future resembles the past. Spot shortages and curtailments of certain hydrocarbon products may occur for short periods, but on the average, no serious shortages may occur. Given the facts listed earlier and this scenario, the following arguments can be made concerning the evolution of the ethylene industry.

1. The price of feedstocks will generally increase in absolute terms, but large increases will not ccur. Natural gas prices will also increase and could reach present LNG prices of around $4.00/million Btu.

2. Because of spot shortages, the ethylene manufacturer will have to prepare himself to handle a variety of feedstocks; the heavier fraction will

have to be used. Cracking of these fractions in the presence of hydrogen and precracking and hydrogenation will be actively pursued to increase ethylene yield. Direct cracking of crude oil will be commercialized, as this process has a great deal of feedstock flexibility.

3. The increased price of natural gas would tend to increase its availability as a result of increased exploration and possible development of the natural gas in the geopressured aquifers.

4. The competition for naphtha by gasoline manufacturers will cause an increase in the price of naphtha and, consequently, this will improve the economics of the ACR.

5. In the event of oil shortages resulting from disruption of production in some of the OPEC countries, such as strikes similar to the one that took place in Iran in 1979 and reduced that country's oil production to a million barrels per day, the effect would depend on the length of the strike and the reduction in output. Short disruptions should have little effect. Small decreases in output could possibly be made up by other oil producers. For disruptions of weeks, the U.S. would have to resort to its stored fossil fuels. However, for long-term disruptions, problems similar to those experienced during the 1974 Arab embargo would be expected. The oil from Mexico, although significant, would not replace large losses in OPEC output of say more than three million barrels per day. However, in the U.S., even with long-term disruptions the petrochemical industry would still have sufficient fossil fuels allocated to it to satisfy its needs. On the other hand, the demand for petrochemicals will definitely be affected, as the total world economy will slow down and cause decreased demand for feedstocks.

6. See No. 5, page 20.

7. See No. 6, page 20.

How plausible is either of these scenarios? Can the present oil glut persist? Will natural gas become more abundant? It was predicted in 1974 that severe oil shortages would be experienced by the mid-1980s. However, a number of factors have since emerged to indicate that they should not occur until possibly the early 1990s. These factors include:

- energy conservation programs
- slowdown of the economies of all the countries of the world
- new oil and gas finds

One may be tempted to add "development of alternate technology" to the above list. However, new energy technologies have had little effect on the demand for fossil fuels. The U.S. does have aggressive plans to commercialize new technologies, such as coal gasification and liquefaction, solar, biomass and wind. But economically it is apparent that conventional fossil fuels have no competition, barring dramatic breakthroughs and massive government subsidies.

When one considers all potential fossil fuel reserves available at prices less than, or equal to, those of alternate technologies, it is apparent that these reserves are enormous. For example, in the Western Hemisphere alone, the oil recoverable from tertiary recovery is sufficient to provide feedstock for the petrochemical industry for 25 years, and the tar sands could supply the industry almost indefinitely. However, to maintain an oil glut, it is necessary to maintain or decrease its present demand growth rate. This can be achieved by developing the alternate fossil fuel—natural gas. In most industrial applications, natural gas is the preferred fuel and is actually worth more than oil because it requires lower capital investment.

The price of natural gas in the U.S. has been permitted to increase to $2/million Btu, a considerable increase over the previous regulated price. It is expected that it would result in renewed interest in exploration in the U.S. As an example, in 1974–1975, Canada appeared to be running short of gas; consequently, its price was increased and all exports of natural gas to the U.S. were discontinued. The increase in gas prices initiated a flurry of exploration. As a result, by 1977 2.4 trillion cubic feet of gas were added to the Canadian proven reserves. Canada then resumed its exports of natural gas to the United States.

Whether this situation will be duplicated in the United States is not yet known; however, U.S. natural gas officials have predicted that natural gas will no longer be in short supply. It would be economically feasible to develop the natural gas reserves in the geopressured aquifers if the price of natural gas were permitted to increase to $4/million Btu (this equates to an oil price of $23/bbl for the same heating value). Technological advances leading to a decreased cost of developing these resources would also encourage their development. Extensive development of these resources would start a new era of energy surplus.

Similarly, the development of the Canadian Cold Lake heavy oils and that portion of the Athabasca tar sands that can be processed by surface mining could add 66 billion barrels to the world's recoverable oil reserves. The technology for developing the reserves is now available and, at today's world oil price, oil from these reserves is economical. Canada has modest plans to produce 0.5 million bbl/day by 1995. The Canadian National Energy Board estimates that under accelerated development conditions, this production could be nearly doubled. Venezuela has similar plans for its tar sands deposits in the Orinoco Belt.

Another expected source of oil is the development proposed in Third World countries with financing from the World Bank. The World Bank estimates that 4 million bbl/day will be available by 1990 as a result of this program.

It appears that both scenarios may be evident in the future. The world has the reserves to continue its oil glut and start a natural gas glut. But energy development programs have long lead times and, if they are not started early enough, we will certainly experience the predicted shortages. Action taken by world governments in the next few years will be critical in determining whether we will have gluts or shortages.

The ethylene industry should not experience feedstock shortages if it continues to be prepared to accept a range of feedstocks; to develop preconditioning technology or hydrogenate the heavier fractions; and to develop new processes to crack crude directly that can accept a range of fractions.

With the present availability of overcapacity, soft market, decreasing demands and enormous capital investments required for major modifications, coal as a feedstock will remain only a possibility. Even if demand were to increase and require new capacity, it is unlikely that coal will make a significant contribution. Biomass, on the other hand, would be the chosen route for the developing countries, but its contribution to world ethylene production will be insignificant.

An optimistic estimate by advocates of the share of petrochemicals that would be manufactured from coal and biomass by the year 2000 amounts to 10–15%.

CHAPTER 4

PROJECTIONS AND USES

INTRODUCTION

Consumption of ethylene is more than all the other olefins combined. U.S. production of ethylene reached 25 billion lb/yr (11.3 million metric tons/yr) in 1977 and is projected to approach 35 billion lb/yr (15.9 million metric tons/yr) in 1982. In response to this, U.S. petroleum refining and petrochemical companies have spent some $2 billion on new capacity since 1974 and are expected to spend another $3–4 billion on new plants now under construction.

Ethylene is one of the primary building blocks for the manufacture of chemicals for plastics, textiles, paper, solvents, dyes, food additives, pesticides and pharmaceuticals, among others (Figure 4-1). In the United States, ethylene consumption is dominated by the production of five large-volume chemicals:

1. *polyethylene*, both low density and high density, the largest volume commodity plastic, now approaching 10 billion lb/yr (4.5 million metric tons/yr);

2. *ethylene oxide* for surfactants, polyester, antifreeze and fiber-grade ethylene glycol, and other solvents, now at about 5 billion lb/yr (2.26 million metric tons/yr);

3. *ethylene dichloride* for polyvinylchloride (PVC) manufacture, now more than 3 billion lb/yr (1.4 million metric tons/yr);

4. *ethyl benzene* for polystyrene manufacture, about 2 billion lb/yr (0.9 million metric tons/yr); and

5. *ethanol* for a host of solvents and chemical manufacturing operations, now at about 1.5 billion lb/yr (0.68 million metric tons/yr).

Ethylene is generally produced by cracking gas and oil products, as will

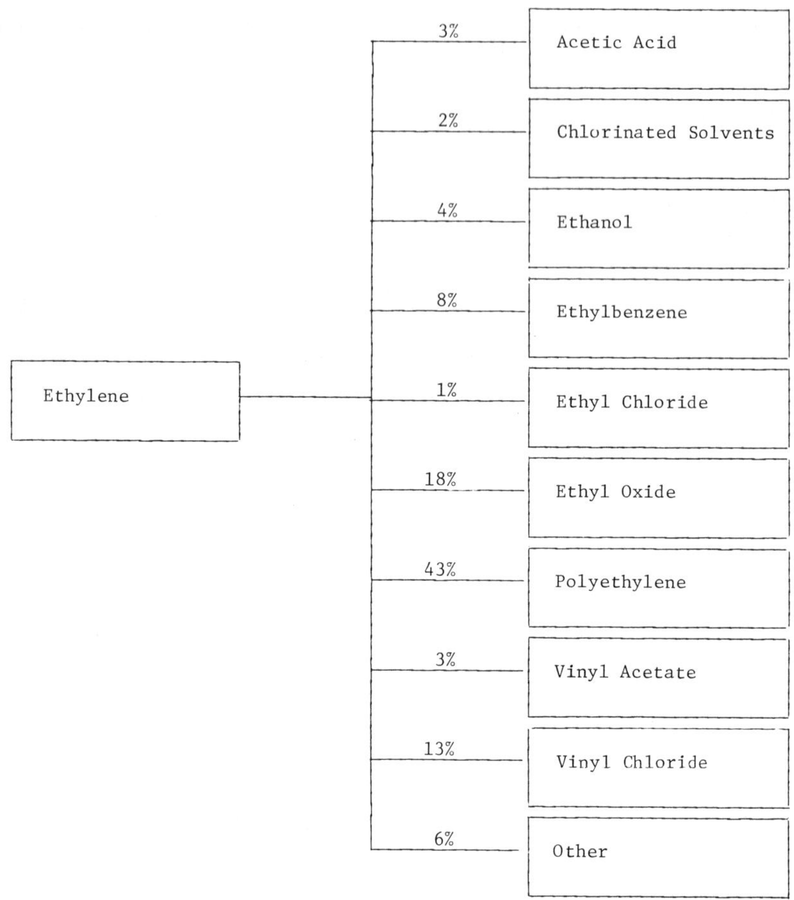

Figure 4-1. End uses of ethylene.

be discussed in the following chapters. Ethylene can also be produced from ethanol by catalytic dehydration, with ethanol produced from biomass. Although this method is not widely used at present, it is uniquely advantageous to oil-poor developing nations. It utilizes locally grown products and employs local labor, while at the same time saving hard currency. In particular, Brazil has an ambitious program to increase its ethanol production to provide 4–6 million m^3/yr by 1985. Although the principal use of this ethanol will be to replace gasoline in transportation applications, it could be used to produce ethylene. Brazil has two small ethanol-to-ethylene plants, which are not operating at present because they are uneconomical.

Another chemical that could conceivably compete with ethylene as a petrochemical building block is acetylene, which is manufactured from coal. At one time, acetylene was the principal feedstock used in the production of a number of chemicals (Figure 4-2), including vinyl chloride and vinyl acetate monomers, neoprene, acrylonitrile and acrylic acid. However, with the introduction of the less expensive ethylene-based processes, ethylene replaced acetylene in the production of the two monomers, butadiene in the production of neoprene, and propylene in the production of acrylonitrile and acrylic acid. However, acetylene remains the mainstay of the metal welding and cutting industry. As a result of the shift to ethylene, acetylene demand decreased from a high of 1150 million lb/yr (0.52 million metric tons/yr) in 1965 to the present production level of 400 million lb/yr (0.18 million metric tons/yr). Monochem operates the only plant in the U.S. making vinyl chloride monomer (VCM) from acetylene, and Borden and National Starch are the only U.S. manufacturers of vinyl acetate from acetylene. The principal area of growth of acetylene has been in the production of low-volume products, such as tetrahydrofuran, acrylate esters and acetylinic chemicals.

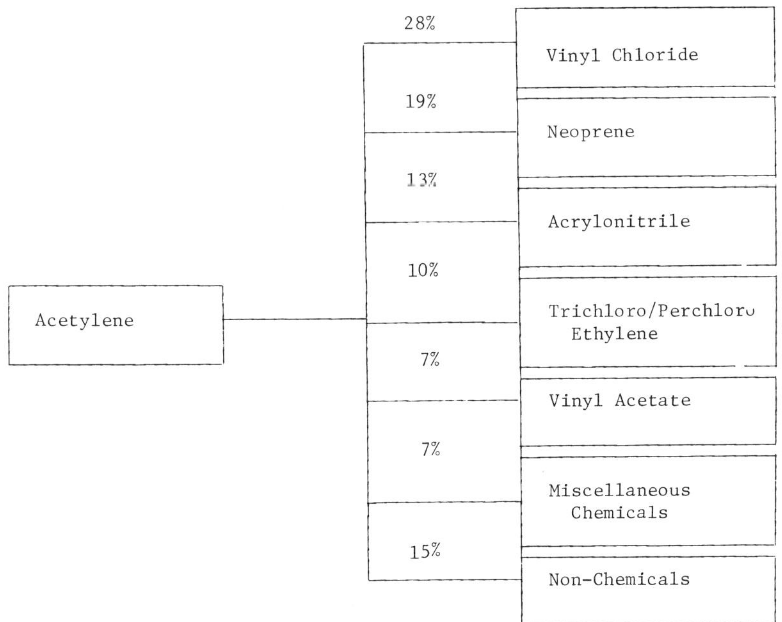

Figure 4-2. End uses of acetylene.

Recent increases in crude and natural gas prices have led to renewed interest in acetylene as a petrochemical building block. Acetylene has an important disadvantage, however; it is a highly unstable and endothermic compound and has a strong tendency to decompose into its elements and to produce large amounts of heat. This creates handling problems, particularly in pipeline transport. As a result, acetylene is generally used or sold locally.

In the U.S., acetylene is produced from coal and natural gas. The coal-based processes include the calcium carbide method, whereby a mixture of calcium oxide and carbon (coke) are heated electrically to between 1800 and 2200° C to form calcium carbide. The calcium carbide is then reacted with water to produce acetylene. Another coal-based process is the AVCO Arc-Coal Process, where pulverized coal is reacted with a hydrogen plasma in a high-temperature (2100–2800° C), high-energy density arc. Acetylene, carbon black and char are produced. The most popular process for acetylene production from natural gas in the U.S. is the partial oxidation process (BASF Process), where a hydrocarbon feedstock (usually (methane) is partially oxidized with pure oxygen. Other methods include the regenerative pyrolysis process (Wulff process) and the electric arc methods. Of the methods that can be utilized to produce acetylene, the AVCO Arc Coal Process is the most economical.

For this discussion, the following processes have been selected:

- ethylene production from crude oil via the naphtha route;
- ethylene production from sugar cane; and
- acetylene production from coal using the AVCO Arc-Coal process.

TRENDS IN U.S. ETHYLENE PRODUCTION

The U.S. ethylene industry has historically depended on the production of natural gas for gas liquids, used for ethylene plant feedstock. Both nonassociated and associated gas productions have been falling for the past several years. Production has exceeded new finds every year since 1968, and domestic production is projected to continue its decline through 1985.

Hydrocarbons heavier than methane have been extracted from natural gas for decades and have become the backbone of the U.S. petrochemical industry. Lower 48 reserves reached a peak of about 290 trillion cubic feet in 1967 and have declined continuously since then. Volumes of gas processed for liquid recovery have followed gas production volumes, with processed volumes peaking in 1972 at 19.2 trillion cubic feet. Furthermore, the average liquid content of wellhead gas has decreased steadily as production has involved deeper wells and offshore gas. Countering this trend somewhat, cryogenic recovery processes have been commercialized, permitting increased ethane and propane recovery. Since the older absorption processes provide full recovery for butanes and natural gasoline, cryogenic technology has no moderating influence on their projected decline in production rates.

Taking into account projections of new gas supplies due, in part, to rising gas prices, Table 4-1 presents a projection of future gas liquids production. Note that this projection portrays a leveling off of supplies at best, and possibly a continuing decline.

Table 4-1. Projected U.S. Gas Liquids Supply (MBPCD)

	1977	1980	1985
Ethane[a]	375	380	420
Propane[b]	510	470	470
Butanes[b]	325	290	280
Natural Gasoline	312	280	270
Total	1522	1420	1440

[a]Some forecasts provide 50% higher supplies under optimistic price assumptions.
[b]Refinery supplies are also highly significant.
Modified from Struth (1979).

By contrast, demand for ethylene is projected to continue to increase at 6–8%/yr rate until the late 1980s (*Oil Gas J.*, November 22, 1976). For example, to supply an additional 20 billion lb/yr of ethylene would require an *increment* of more than 550 MBPCD of ethane, or nearly 800 MBPCD of propane. These increments exceed the total available supply, part of which is currently dedicated to ethylene production. Hence, conversion from gas-liquids to refinery naphtha or gas oil for ethylene production is not a question of feedstock prices or plant economics; it is a necessity to provide access to otherwise unavailable feedstocks. Table 4-2 and Figures 4-3 and 4-4 illustrate the future importance of heavy refinery liquids as a source of ethylene supply; of 12 new world-scale olefins plants on the Gulf Coast, only one is based on natural gas liquids. Furthermore, the increased butadiene production with naphtha/gas oil relative to both ethane and propane cracking will have an important influence on the future U.S. butadiene supply/demand balance. This will have a substantial impact on the current tight market supply, greatly decreasing future import requirements, which would otherwise be needed, and shutting in much of the current dehydrogenation capacity.

These trends will have substantial impacts on product pricing projections. In general, return on equity of ethylene manufacture will drop from

Table 4-2. Heavier Feedstocks Predominate for New U.S. Ethylene Capacity

Onstream	Company Location	Capacity (Million lb/yr)	Feedstock
1977	Arco, Channelview, TX	1300	Gas oil
	Amoco, Alvin, TX	1000	Ethane and naphtha
1978	Texas, Port Arthur, TX	1000	Naphtha and gas oil
	Union Carbide, Taft, LA	700	Naphtha
1979	Dow, Plaquermine, LA	1200	Naphtha
	Phillips, Sweeny, TX	1000	Ethane
	Shell, Deer Park, TX	1500	Naphtha and gas oil
1980 or later[a]	Exxon, Baytown, TX	1200	Naphtha and gas oil
	ICI-Champlin-Soltex, Corpus Christi, TX	1200	Naphtha and gas oil
	Gulf, Gulf Coast	1200	Gas oil
	Conoco-Monsanto, Gulf Coast	900	Gas oil
	Aroc-Du Pont, Gulf Coast	1000+	Gas oil
	Shell, Norco, LA	1500	Gas oil
	Arco, West Coast	1200	Naphtha

Source: *Chemical Engineering*, March 28, 1977.

[a]Only the first two of these appear to be firm; the others are in various stages of feasibility studies.

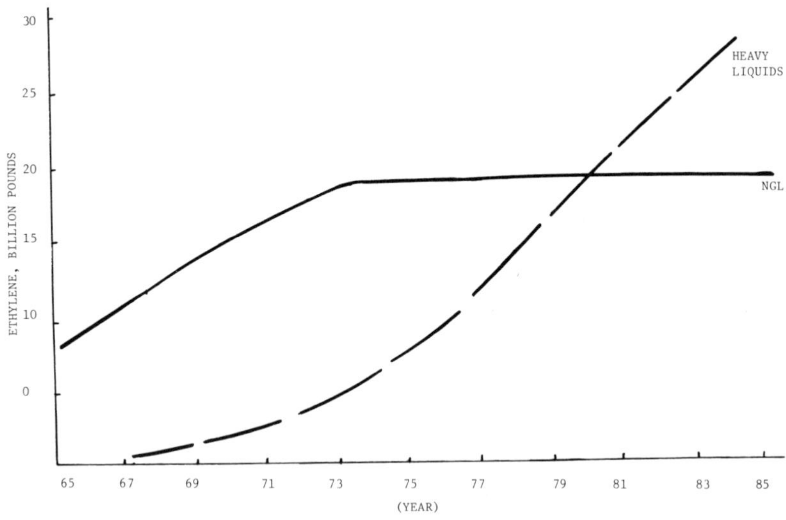

Figure 4-3. Future ethylene feedstock sources.

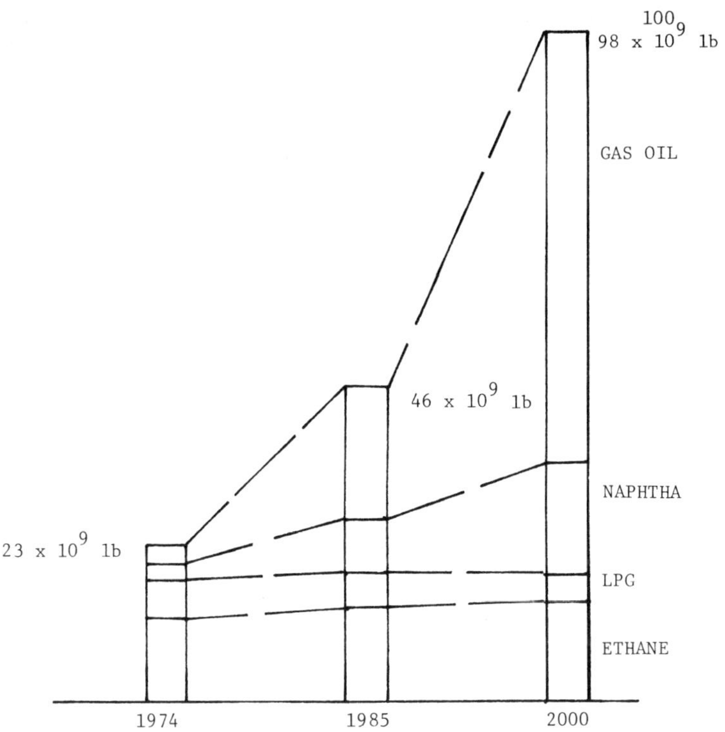

Figure 4-4. Feedstocks for ethylene in the United States.

current levels as the benefits of old plants are used up. The pricing of ethylene from new plants will tend to remain just below replacement cost pricing levels, particularly with the new capacity additions coming onstream.

Historical price allocation formulas will have to be adjusted to reflect the larger coproduct production. Naphtha values will be dictated by alternative uses as reformer feedstock for gasoline production. Ethane values will be dictated by the economics of cryogenic gas liquids recovery plants. Pyrolysis gasoline will be valued relative to unleaded gasoline. Propane will command a premium of 40–50¢/million Btu over distillate fuel oil. Butane's price will be dictated by alkylation economics, with isobutane about 2¢/gal over normal butane. Butadiene has historically been priced based on dehydrogenation economics, but this is expected to shift to a price basis of fuel value plus extraction costs.

Yield

The trend toward use of heavier feedstocks in ethylene plants will mean coproduction of significant quantities of by-products. Gas oils typically produce a lower ethylene yield than other feedstocks (Table 4-3). These plants have higher investment costs for the same ethylene production. Another important characteristic of gas-oil cracking is a significant production of heavy fuel oil, normally far in excess of fuel oil requirements of ethylene plants.

Table 4-3. Olefin Plant Yields[a] (wt %)

Product	Feedstock Type			
	Ethane[b]	Propane[c]	Naphtha	Gas-Oil
Hydrogen	5.6	1.3	1.2	0.9
Methane	8.2	27.3	16.8	12.7
Ethylene	80.1	45.4	32.8	27.5
Propylene	1.8	15.1	12.7	11.9
Other C_3	0.9	–	0.8	1.1
Butadiene	1.1	1.6	5.1	4.8
Other C_4	0.8	1.0	4.0	3.8
Pyrolysis Gasoline	1.5	8.3	8.4	8.2
Benzene	0.0	0.0	6.0	7.0
Toluene	0.0	0.0	3.5	3.8
C_8 Aromatics	0.0	0.0	3.5	2.6
Fuel Oil	0.0	0.0	5.2	15.7
Total	100.0	100.0	100.0	100.0

[a]Ethane recycled to extinction.
[b]60% once-through conversion.
[c]93% once-through conversion.

Furthermore, with naphtha or gas oil feedstocks, the yield patterns shift substantially from those based on ethane/propane feedstocks. Note from Table 4-3 that the ethylene once-through yield drops from 80% with ethane to 27% with gas oil. A substantial increase in raw materials requirements per pound of ethylene produced is quite evident. This, in turn, results in an increased importance of coproduct values in determining ethylene values. Note, for example, the increased production of aromatics in the pyrolysis gasoline, as well as the increased fuel oil production.

Further trends are evident in Table 4-4, in which the yields are normalized to reflect a fixed production of ethylene from the various feedstocks. Note that the relative production of coproduct olefins from heavy liquids cracking is greatly increased, particularly relative to the ethane case.

Major Trends

1. Ethylene production growth will average 6–8%/yr.

Table 4-4. Relative Olefin Plant Yields (with ethylene = 100)

Product	Ethane	Propane	Naphtha	Gas-Oil
		Feedstock Type		
Hydrogen	7.0	2.9	3.7	3.3
Methane	10.2	60.2	51.2	46.2
Ethylene	100.0	100.0	100.0	100.0
Propylene	2.3	33.3	38.7	43.3
Other C_3	1.1	–	2.4	4.0
Butadiene	1.4	3.5	15.6	17.5
Other C_4	1.0	2.2	12.2	13.8
Pyrolysis Gasoline	1.9	18.3	25.6	29.8
Benzene	0.0	0.0	18.3	25.5
Toluene	0.0	0.0	10.7	13.8
C_8 Aromatics	0.0	0.0	10.7	9.5
Fuel Oil	0.0	0.0	15.9	57.1

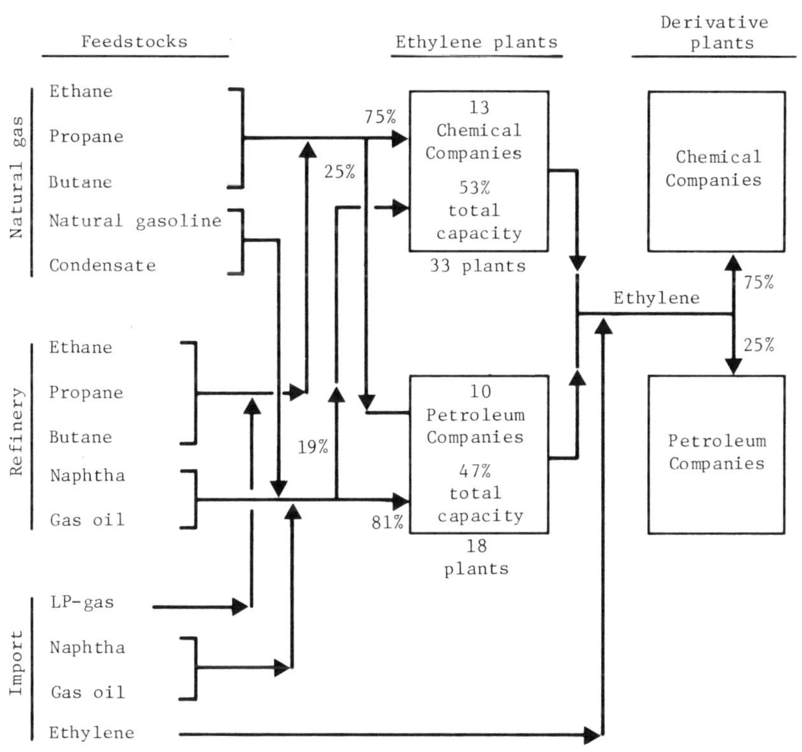

Figure 4-5. U.S. ethylene industry, 1976 (Minet and Tsai, 1977).

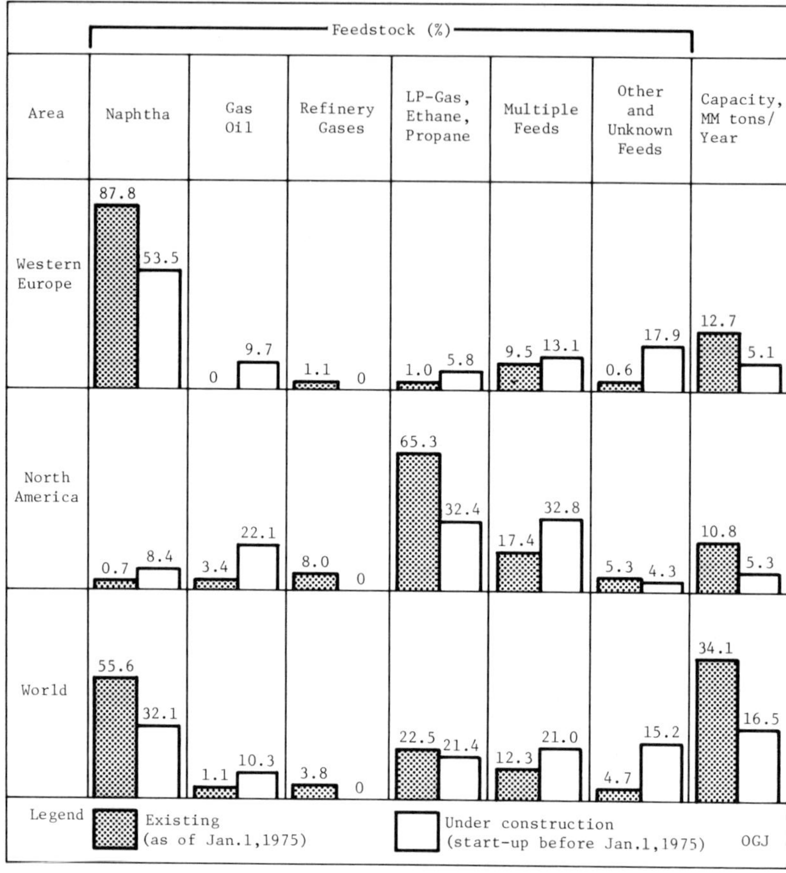

Figure 4-6. World ethylene feedstock and capacity (*Oil Gas J.*, October 18, 1971).

2. Raw material for feedstock price will increase steadily.
3. Joint ventures between chemical and oil companies will be more frequent.
4. The variability and value of by-products will have a major impact on ethylene production and price.
5. Almost the entire growth in ethylene production will come from the heavier feedstocks—gas oil and naphtha.
6. The future use of ethane feedstock will be basically supply limited and relatively insensitive to changes in demand for energy and ethylene.
7. The economic level of use of naphtha and butane as feedstock is highly sensitive to demand for gasoline.
8. Propane will decline in importance as an olefin raw material.
9. Gas oil is the "swing" feedstock.

Table 4-5. European Ethylene Capacity Growth (10^3 metric ton/yr)

	Capacity Onstream	Under Construction or Planned
United Kingdom		
BP Chemicals	775	
ICI	650	
Shell Chemical	200	500, 1981
BP-ICI (joint venture)		500, 1978
Esso Chemical	105	400–600, late 1981
West Germany		
Rheinische Olefinwerke	1200	300, 1978
Veba-Chemie	750	
Erdoelchemie	460	360, 1978
RWE-UK Wesseling	500	
Esso Chimie	450	
Caltex	320	
Deutsche Marathon Petroleum	205	20, 1978
Chemische Werke Hüls	40	
Netherlands		
Dow Chemical	900	
Shell	650	
Gulf	300	
DSM	500	450, 1978
Italy		
Montedison	960	250, 1980
Anic	340	
Montedison-Anic		550, 1979
Soc. Italiana Resine	305	
Rumianca	65	300
Solvay	55	
France		
Naphtachimie	620	
CdF-Chimie	440	225, 1979
ATO Chimie	375	
ELF-Aquitaine	100	
Shell Chimie	100	350, in discussion
Die. Francaise des Petroles		
Kuhimann, Solvay (joint venture)	60	
Esso Chimie	300	
Belguim		
Petrochim	500	Double in 1980s

Source: *Chemical Week,* March 30, 1977.

THE WORLD ETHYLENE INDUSTRY

In 1976, 51 U.S. ethylene plants were owned by 13 chemical and 10 petroleum companies (Figure 4-5). Chemical companies made up 53% of the capacity and petroleum companies 47%. Chemical companies consumed 75% of the gas feed (ethylene, propane and butane), and petroleum companies used 81% of the liquid feed (naphtha, gas oil and condensate) (Minct and Tsai, 1977).

An analysis of world feedstock (*Oil Gas J.*, October 18, 1976) for ethylene indicates some dramatic differences between North America and Western Europe, as well as some striking similarities (Figure 4-6). For example:

- Western Europe's ethylene is primarily based on naphtha; while in the U.S., LP gas, ethylene and propane were the major sources in the 1975 base year.
- Ethane and LP gas are replacing naphtha as the prime source of ethylene in Western Europe. In the U.S., naphtha will play a larger role in the future (Table 4-2).
- Flexibility and multifeed is the order of the day for new capacity.

Recent information of European ethylene capacity is provided in Table 4-5 (*Chem. Wk.* May 11, 1977).

- Most of the future ethylene plants in the U.S. will be built by oil companies or by joint ventures with chemical companies. A joint venture between an oil firm and a chemical company gives the latter a secure source of feedstock and an outlet for unwanted by-products.
- A potential problem for the olefin plant operating on heavy feedstocks is its susceptibility to variation in by-product values.

CHAPTER 5

ALTERNATIVE FEEDSTOCKS

INTRODUCTION

This chapter examines the economics of the production of ethylene from crude oil and biomass and compares these to the economics of producing acetylene from coal. To compare the cost of ethylene with that of acetylene requires a number of assumptions about the utilization of the two chemicals. For the purposes of this chapter, it was felt that a direct comparison of the price of ethylene made from crude oil, biomass and coal would be useful. It is possible to manufacture acetylene from ethylene or from naphtha; it is also possible to manufacture ethylene from acetylene and to manufacture ethylene directly from coal, but neither method is recommended because of the poor economics of each. Hence, consideration of these processes will not yield a meaningful comparison between the cost of ethylene and that of acetylene. To obtain a meaningful comparison it was decided to carry out the manufacturing procedure one more step and produce a chemical that is currently being produced from ethylene and acetylene. Vinyl chloride monomer (VCM) is an important chemical, which ranks ninth among the organic and inorganic chemicals produced in the U.S. and is manufactured from ethylene and acetylene. The use of VCM then will lend a measure of realism to the comparison of the cost of ethylene to that of acetylene.

The approach, as shown in Figure 5-1, is as follows:

1. The cost of ethylene is determined from crude oil cracking.
2. The cost of ethylene is determined from the biomass process.
3. The cost of acetylene is determined for the AVCO Arc-coal process.

The cost of VCM is then determined as a function of the cost of ethylene

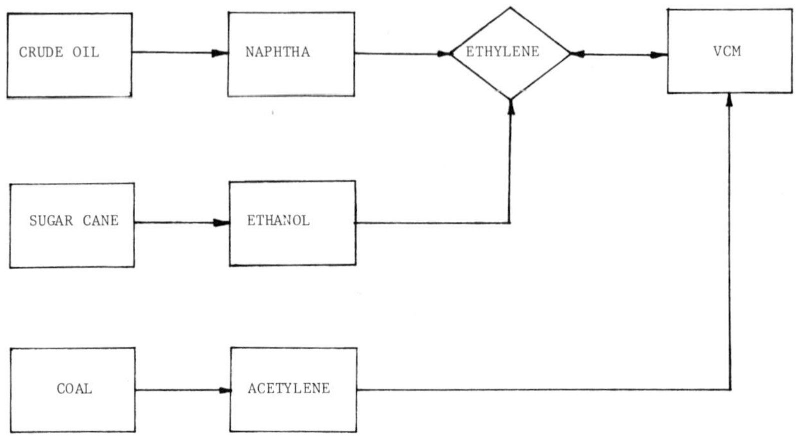

Figure 5-1. Steps utilized in arriving at the cost of ethylene from the three feedstocks.

and as a function of the cost of acetylene. This is used to arrive at an "equivalent ethylene" price for acetylene.

INDEPENDENT VARIABLES

Given the present method of producing ethylene from crude oil, which is to crack the crude to produce naphtha and then crack the naphtha to produce ethylene, it is apparent that the cost of ethylene from crude should have a strong relationship to the cost of the crude oil feedstock. In addition to the feedstock, other forms of energy products are used in the production of ethylene: these include steam, fuel and electricity. As shown in Table 5-1, the largest portion of the energy is for fuel. Hence, it can be assumed that all manufacturing power requirements will vary with the cost of crude oil. Thus, in the case of ethylene from crude, the cost of crude oil is taken as an independent variable.

With biomass, sugar cane is fermented to ethanol, which is dehydrated to produce ethylene. It can be assumed that the price of the ethylene from biomass should depend on the price of sugar cane only. Sugar cane must be fertilized with products derived from crude oil, and oil-consuming equipment must be utilized to produce the sugar cane. This would imply that the price of sugar cane should be a function of crude oil prices. There are factors, however, that have much greater effect on the cost of sugar cane than the cost of oil: these include weather, sugar demand and ethanol demand. Any of these factors could have a greater effect on the cost of sugar cane than the cost of oil and are independent of the cost of oil. Hence,

Table 5.1 Annual Olefin Plant Operating Costs[a]

Cost Item	Unit Cost	Unit Usage	Total Cost ($ million)
Power, 10^6 kWh	23,000	50	1,150
Fuel, million Btu	2.05[b]	15,000,000	30,750
Steam, million lb	3,080[c]	1,000	3,080
Catalyst and Chemicals, $	1	1,700,000	1,700
Cooling Water, million gallons	15	60,000	900
Boiler Feed Water, million gallons	2,000	80	160
Labor	[d]	–	2,514
Maintenance	4%[e]	–	8,640
Taxes and Insurance	2.5%[e]	–	5,400
Total			54,294

[a]Excludes interest and depreciation.

[b]At $21/bbl crude, with 5.85 x 10^6 Btu/bbl. Will be adjusted with new crude prices. Actual U.S. fuel prices depend on domestic crude usage, as well as on pricing and gas availability and pricing.

[c]Fuel price times 1500 to reflect increased cost of steam with increasing energy prices.

[d]Twelve men per shift, 4.6 shifts, $12/hr plus 20 supervisors at $29,000/yr.

[e]Based on onsite plus offsite investment.

for our purposes, the cost of sugar cane is considered an independent variable.

When it comes to the price of acetylene, the AVCO Arc-Coal Process was chosen. In this process, coal and electricity are used in the production of the acetylene. A review of coal and oil prices indicates a correlation between them. When the price of oil went up, the price of coal followed. It should be noted, however, that the cost to produce a ton of coal changed very little; rather, the price of coal climbed in response to increased demand. In the United States there are large coal reserves, but environmental considerations have prevented a massive shift to coal by energy consumers, especially the utilities. Rather than be entangled in some assumptions about oil/coal price relationships, we have taken the price of coal as an independent variable.

Similarly, it can be shown that the price of electricity should be related to the price of crude oil. Electricity may be produced from crude oil, natural gas, coal, hydro, nuclear, or from a combination of these. In the case of the AVCO process, it appears economical to have the electricity generated using coal or the char by-product. In this case, the cost of electricity would be directly related to the cost of coal. For this study, the price of electricity was initially considered an independent variable; later it is assumed to be related to the price of coal.

DEFINITION OF PROCESSES USED

Ethylene from Crude Oil

Ethylene is normally produced by pyrolysis of hydrocarbons. The pyrolysis is usually conducted at high temperature (700–800° C), short residence times (approximately 1 second) and low hydrocarbon partial pressures. Not all hydrocarbons pyrolyze with equal ease; cracking yields vary considerably with the type of hydrocarbon feedstock and cracking severity.

Here, the production of ethylene from crude is assumed to be achieved according to the conventional two-step process: (1) crude oil is cracked to yield naphtha; and (2) naphtha is pyrolized to produce ethylene. The crude oil feedstock to the naphtha cracker is assumed to be Arabian light, which results in naphtha with 55° API oil gravity. Severe pyrolysis conditions will be assumed in the naphtha pyrolyzer, resulting in the product mix shown in Table 5-2. Detailed descriptions of the crude oil and naphtha cracking processes are well documented in the literature.

Table 5-2. Olefin Plant Yields Naphtha Feedstock

Product	Yield (wt %)	78 Price[a] (¢/lb, 1978)	Credits
Hydrogen	1.2	20.5	0.0[b]
Methane	16.8	4.5	0.76
Ethylene	32.8	x	0.328x
Propylene	12.7	10.0	1.27
Other C_3	0.8	7.3	0.06
Butadiene	5.1	21.0	1.07
Other C_4	4.0	8.0	0.32
Pyrolysis Gasoline[c]	21.4	6.9	1.48
Fuel Oil	5.2	4.1	0.21
	100.0		5.15[d]

[a]1976 price inflated @ 7.5%.
[b]Flared locally.
[c]Including BTX.
[d]The by-product credits are then 5.15/0.328 = 15.70¢/lb of ethylene.

The product mix shown in Table 5-2 is the one that will be used throughout this study. Except for hydrogen, the by-products are valuable, and credits for their value should be taken into account. The hydrogen produced in this process is normally mixed with impurities and is of little economic value; it is generally flared onsite.

The naphtha cracking plant size utilized for this study has a capacity of 1 billion lb/yr of ethylene production, which is a world-size plant.

Ethylene from Sugar Cane

In situations where fossil energy materials are not available and biomass materials such as sugar cane, beet sugar and cassava are readily produced, it may be feasible to produce ethylene from ethanol obtained from the fermentation of biomass materials. The reaction takes place at 340–395° C over a specially treated alumina catalyst in fixed-bed reactors. The catalyst is periodically regenerated to remove carbon with an air-stream mixture. In this process, the conversion rate from ethanol to ethylene is around 95%.

Conversion of the sugar cane extract to ethylene proceeds through two main chemical reactions:

$$C_{12}\ H_{22}\ O_{11} + H_2O \xrightarrow{\text{Yeast}} 4C_2\ H_5\ OH + 4CO_2$$

$$\underset{\text{Sugar}}{\phantom{C_{12}\ H_{22}\ O_{11}}} \qquad\qquad \underset{\text{Ethanol}}{}$$

$$C_2\ H_5\ OH \xrightarrow{\text{Catalyst}} \underset{\text{Ethylene}}{C_2\ H_4 + H_2O}$$

In the fermentation/distilling operation in which ethanol is produced, there is a large amount of bagasse by-product. This bagasse is more than sufficient to provide all the heat energy required by the distillery. Some electric power is also required to operate electric motors. Excess bagasse is available and, in a combined distillery/ethanol to ethylene plant, can be used to provide the heat required for the ethylene plant.

For the purposes of this chapter, two separate plants are assumed; one for the production of ethanol from sugar cane and one for the production of ethylene from ethanol.

For ethanol production, the processing facility would consist of the elements shown in Figure 5-2. In this case, few economies of scale could be achieved by making these elements larger, as the size of the elements is related to the fiber content of the sugar cane, rather than to its sugar content. After harvesting, the sugar cane is unloaded and shredded in three successive knife shredders. Knife shredders are chosen over busters and fiberizers because of their higher energy efficiency. This operation is followed by the juice extraction operation, in which the prepared sugar cane is passed through four three-roller mills in sequence. Process water is required for this step, and the amount of water depends on the quality of the sugar cane. If the sugar cane tops and leaves are included, the water requirements are different. The juice is then raw material for the fermentation process. The major disadvantage of using the juice without further

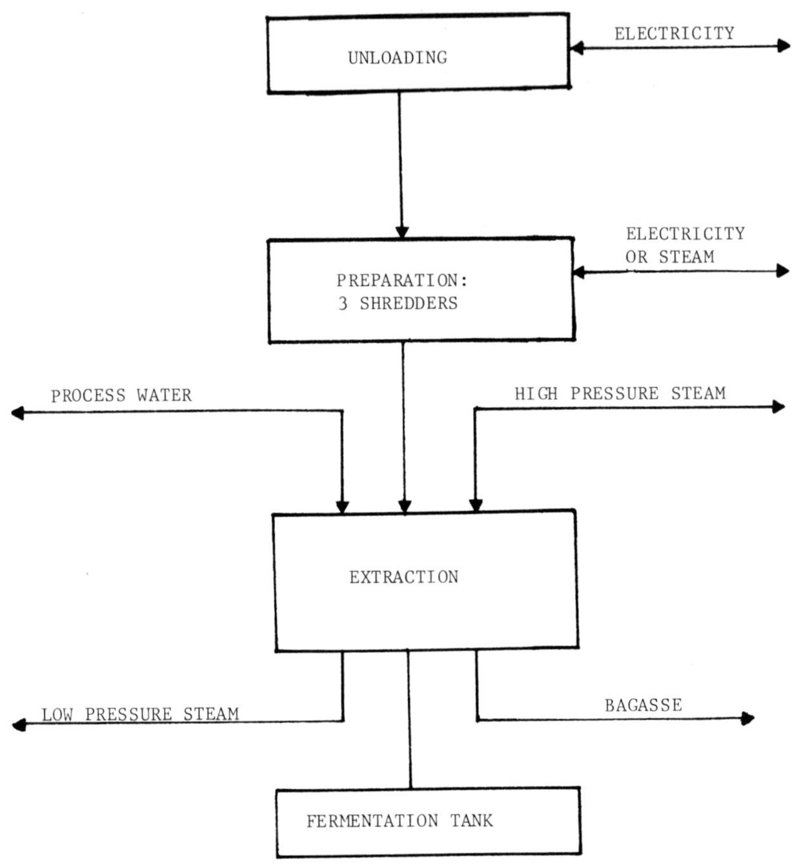

Figure 5-2. Ethanol from sugar cane.

processing is that it is subject to spoilage unless it is concentrated roughly to the consistency of molasses. Alternatives to concentration, such as heat sterilization, chemical sterilization and ultrafiltration to remove microbes, are effective in preventing spoilage; however, the quantity of liquids to be stored would be great, resulting in high storage costs. Storage costs are especially important in this case because the sugar cane harvesting season is normally about three months in temperate regions and it is necessary to operate the plant at least 330 days/yr to make it economical.

Conversion to molasses is therefore desirable from an economic standpoint. The resulting molasses is fermented in large tanks under controlled temperature conditions. Yeast is added to promote the fermentation. After the fermentation step, the product is distilled to remove the water and obtain high-purity (94–96° GL) ethanol.

Although for our purposes ethanol is assumed to be produced from sugar cane only, it can actually be produced from a number of other biomass materials. Medville et al. (1978) have thoroughly examined the alternate biomass raw materials and the cost of production of ethanol from these other materials.

The ethanol is then used as a feedstock to the ethylene plant. A diagram of the ethylene plant is shown in Figure 5-3. A world-size plant has a capacity of 60,000 metric tons/yr and the plant utilized for this study is that described by CE Lummus (Tsao et al., 1978). Ethanol is pumped to system pressure and passed over the catalyst. The dehydration reaction is endo-

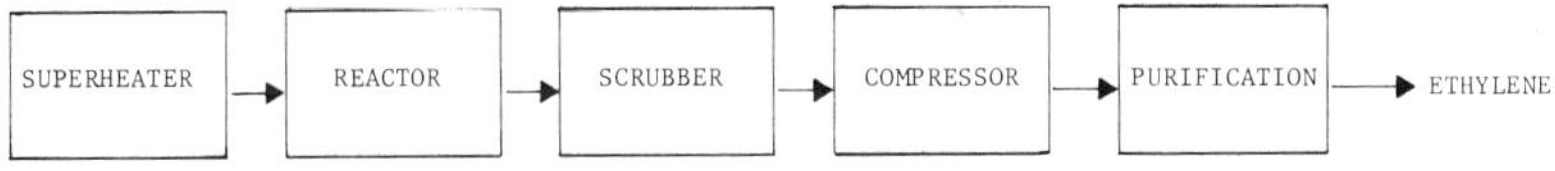

Figure 5-3. Diagram of the ethylene from ethanol plant.

thermic and requires net heat input. The operating temperature is critical, as high temperatures will result in the formation of aldehydes and low temperatures in the formation of ethers. The reactor output stream, which consists mainly of ethylene and reaction product water, is quenched by direct contact with water and a solution of unreacted ethanol. The cooling fluid will normally contain some by-product impurities (e.g., ethers and aldehydes). Following the quench vessel, the cooled vapor is treated to remove the carbon dioxide by direct contact with a countercurrent of a dilute caustic stream. The resulting "wet" ethylene is then compressed, cooled and passed through an activated carbon bed to remove the heavy ends, such as C_4. This operation is followed by another adsorbtion process utilizing a desiccant to remove the water. The output ethylene stream is better than 99% pure.

Acetylene from Coal (Arc Process)

Traditionally, petrochemical producers seeking alternate production pathways have viewed synthesis of gas as the most likely use of coal.

Acetylene production is another alternative and could be produced from coal through two routes: (1) from calcium carbide, or (2) via the AVCO consumable arc process.

For our purposes, the AVCO consumable carbon anode arc process is evaluated as the method for producing acetylene as a petrochemical building block. A diagram of the reactor is shown in Figure 5-4. The anode consists of a hollow water-cooled copper tube, through which crushed coal is fed. A high-intensity arc is sustained between the cathode and coal in the anode in a hydrogen atmosphere. The gasification reaction takes place in

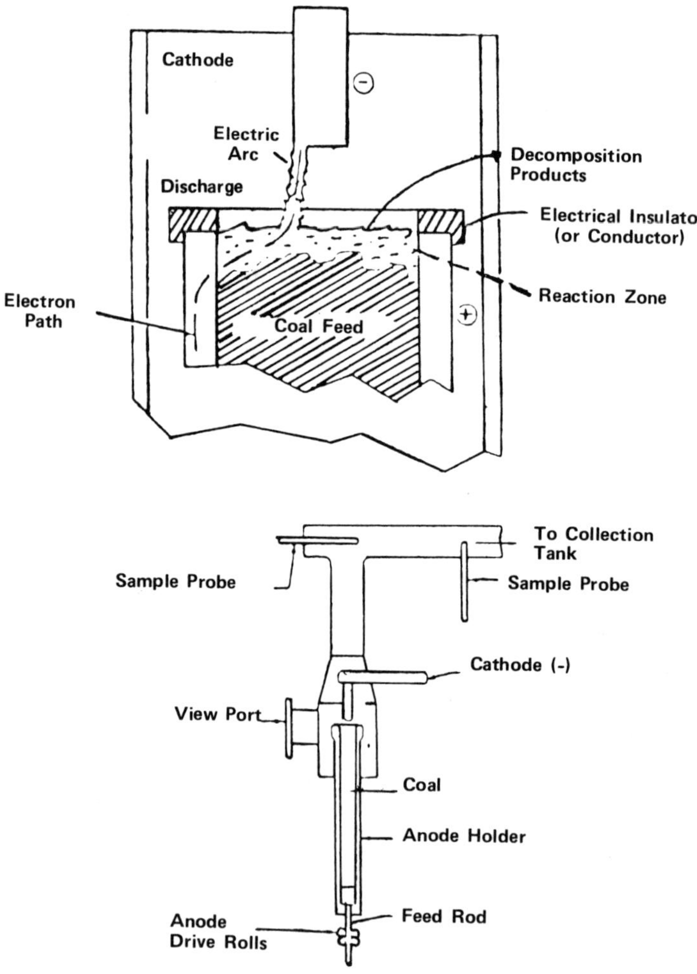

Figure 5-4. Schematic of coal conversion of arc process and coal reactor (Dravo, 1978).

the hydrogen plasma between the cathode and anode. Energy is transferred to the coal through two mechanisms: by the Joule heating and by the electrons colliding with the surface of the carbon particles. The materials leaving the reactor consist of gas products, carbon black, char agglomerates and some unconverted coal. A major research and development goal is the development of a process to separate char from carbon black. The gas products include acetylene, H_2, CH_4, HCN, CO, CS_2, H_2S, C_2H_4 and C_3H_6.

There are five steps in the complete process, as shown in Figure 5-5:

1. Coal is prepared to a 2% moisture.
2. Reaction takes place in a water-cooled jacketed reactor into which coal is continuously screw-fed and arc-burned at 2150° C. Hot gas overhead is quenched with recycle gas to 450° C to prevent acetylene decomposition.
3. Acetylene is compressed, cooled, dried and separated. Char, dropping from the reactor, is cooled by water to generate by-product steam, which is used to power the electricity generation to energize the arc.
4. Inert gas from acetylene separation bearing CO and H_2 is used to pneumatically transport char to the power plant.
5. Acetylene separation involves acid gas removal using a soda ash solution that scrubs out the HCN by-product, which is recovered from the soda ash solution in a stripping column. Acetylene overhead gas from the soda ash solution scrubber is recovered by ammonia solution scrubbing in a secondary column, then scrubbed with methanol to remove CS_2 traces, and finally scrubbed with a copper-ammonia-formate solution to remove CO.

This process has not been commercialized; therefore, economic evaluation is based on the hypothesis of a world-scale plant producing 300 million lb/yr of acetylene.

Because acetylene is not easily transported, the arc process acetylene plant would have to be located adjacent to the VCM plant. A 715 million lb/yr vinyl chloride plant would be required in addition to the adjacent acetylene plant, as well as a nearby chlorine-producing unit. The chlorine plant would be on the order of 400 million pounds of chlorine and consume approximately 1000 tons/day of salt. Therefore, the acetylene-vinyl chloride-chlorine complex would have to be adjacent to a source of low-cost salt, as well as very cheap power and readily available low-cost coal. The latter two requirements imply that an adjacent, huge coal-fired power plant would be part of the complex.

Acetylene to Vinyl Chloride Monomer

Although the purpose of this chapter is to compare the cost of production of ethylene by the three routes—crude oil, biomass and coal—direct production of ethylene from coal is not economical. Acetylene is

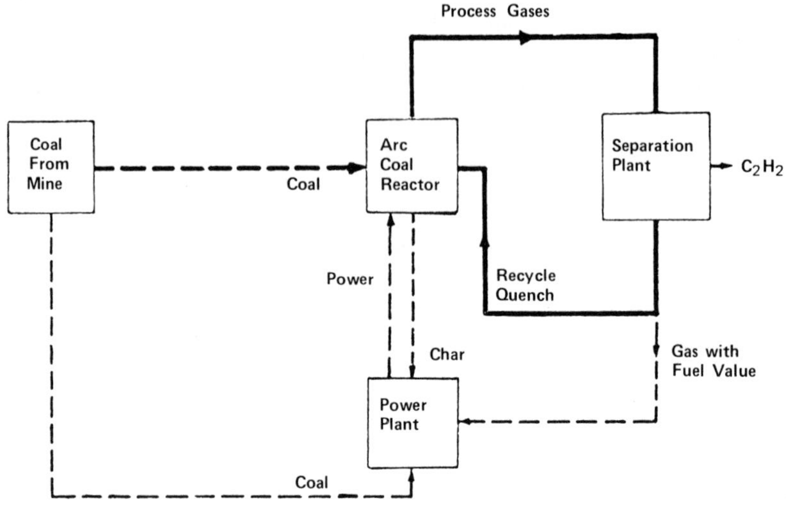

Figure 5-5. Diagram of an arc-coal plant (Dravo, 1978).

normally the basic feedstock produced from coal, and a comparison of the cost of production of acetylene to the cost of production of ethylene is meaningless. To obtain a meaningful basis for comparison, it was decided to introduce an additional step to produce a chemical that is normally produced from ethylene *and* acetylene. VCM is an important chemical now produced in large quantities from ethylene and acetylene. The chemical reaction is:

$$C_2 H_2 + HCl \xrightarrow{\text{Catalyst}} CH_2 CHCl$$

$$\text{Acetylene} \qquad\qquad \text{VCM}$$

Typically, the acetylene is dried by passage through a drying tower containing a desiccant and is mixed with dried hydrogen chloride in a mixing chamber containing active charcoal. An important function of the charcoal is to absorb any chlorine that may be present in the hydrogen chloride and could result in the formation of chloroacetylene, which is explosive. The mixture is then passed through the catalyst (e.g., mercuric chloride adsorbed on active charcoal), where the two chemicals are reacted. Reaction temperature begins at 80° C and is increased to 250° C as the catalyst deteriorates. The output of the reactor is scrubbed to partially remove some of the impurities. Acetylene conversion in this case can be as high as 99%.

The hydrogen chloride required is usually obtained from other processes in the petrochemical complex and has negligible value.

Ethylene to Vinyl Chloride Monomer

Ethylene can be chlorinated directly to VCM at 300–500° C without the presence of a catalyst. However, it has been shown (Yu et al., 1962) that this reaction results in considerable destruction of ethylene and production of a large number of unwanted chlorinated hydrocarbons. The preferred method of VCM manufacture is through the balanced oxychlorination process, which involves three principal steps (Figure 5-6).

The first step involves the production of ethylene dichloride (1,2-dichloroethane) by direct chlorination of ethylene at relatively moderate temperatures (15–135° C) in the presence of a catalyst such as ethylene bromide, iron chloride or aluminum chloride, as follows:

$$C_2\ H_4\ +\ Cl_2 \xrightarrow{\text{Catalyst}} Cl\ CH_2\ CH_2\ Cl$$

The second step is the pyrolysis of the ethylene dichloride at high temperature (480–510° C) and pressure (50 psig) over a catalyst such as pumice or charcoal. The products are VCM and hydrogen chloride, and the conversion is 50–60%, which necessitates the recycling of the unreacted ethylene dichloride to the pyrolysis unit. The reaction is as follows:

$$Cl\ CH_2\ CH_2\ Cl \xrightarrow{\text{Catalyst}} CH_2\ CH\ Cl\ +\ HCl$$

Another step is the oxychlorination of ethylene to form ethylene dichloride at temperatures between 330 and 420° C. This step utilizes the hydrogen chloride produced in the previous reaction and requires oxygen as follows:

$$2\ C_2\ H_4\ +\ 4HCl\ +\ O_2 \rightarrow 2\ Cl\ CH_2\ CH_2\ Cl\ +\ 2\ H_2O$$

The overall reaction for the balanced oxychlorination process is:

$$4\ C_2H_4\ +\ 2\ Cl_2\ +\ O_2 \rightarrow 4\ CH_2\ CHCl\ +\ 2\ H_2O$$

The overall conversion of the input ethylene into VCM is 90% for this process. Generally, a captive chlorine plant is utilized, and chlorine is obtained by electrolysis of salt.

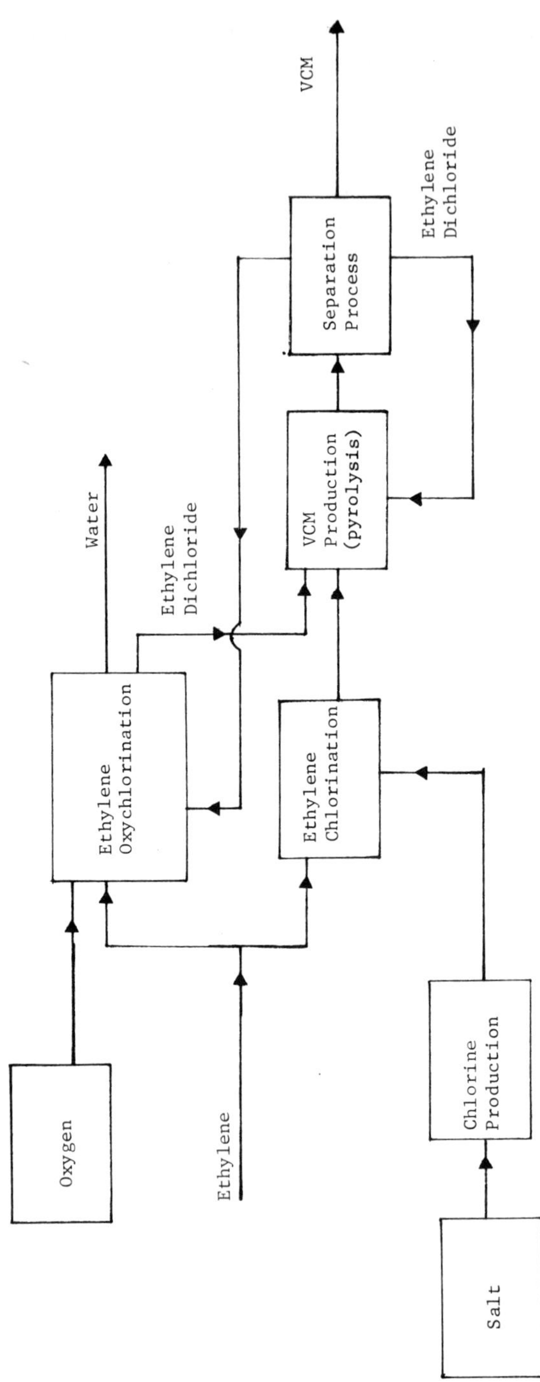

Figure 5-6. The balanced oxychlorination process for the production of VCM from ethylene.

CHAPTER 6

COST COMPARISONS

GENERAL PROCEDURE

The cost allocation process for each chemical can be represented diagrammatically, as in Figure 6-1. The manufacturing process is represented by a box in which feedstock, labor and ancillary materials enter and the desired product and coproducts exit. Costs associated with production, both out-of-pocket expenses and capital amortization, must be recovered over the expected life of the plant. The box in Figure 6-1 also represents a MITRE/Metrek computer model (described in Chapter 9), which determines the revenue requirements for each year. The costs are then leveled to produce a uniform cost per unit product. The cost of the input stream was then varied and the resulting cost of the product determined.

Realistic cost estimates of the alternative chemical products require complete and consistent estimates of the capital investment required. Capital investment should include the cost to construct the plant; the cost of all ancillary equipment, such as storage facilities, roads, steam generators, desulfurizing equipment and pollution control equipment; and the cost of capital, taxes and working capital. Certain simplifying assumptions had to be made to remain within the scope of the study. Some of these assumptions were required to allow the study to be performed within a reasonable time, while others were dictated by the lack of reliable data because of the newness of one of the processes under consideration, i.e., the arc-coal process.

The arc-coal process is experimental, with no model plants available to determine some of the required data. Available data for capital and operating costs are preliminary engineering estimates. Since this process requires a coal source and significant amounts of electricity, it has been

49

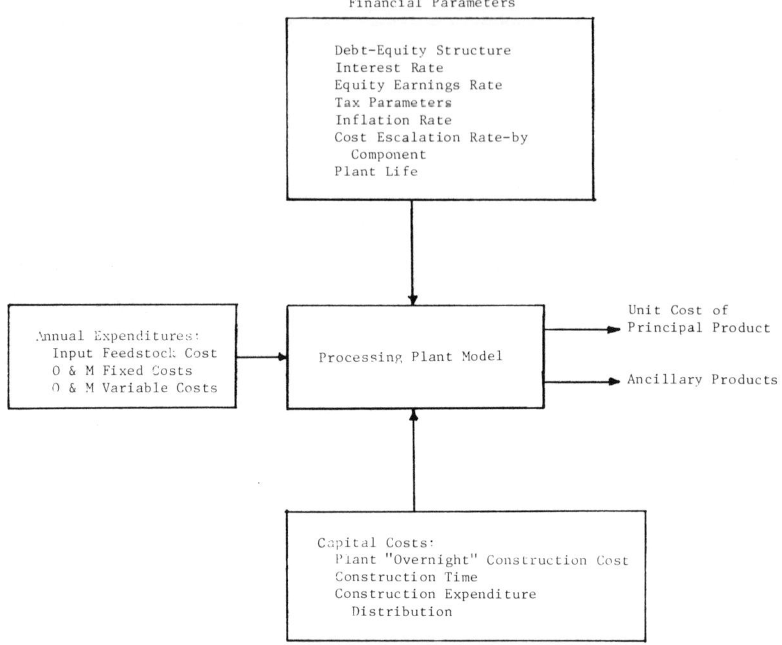

Figure 6-1. Illustrative chemical process cost flow.

proposed to construct the acetylene plant at the mine mouth as part of a large carbochemical complex, which would include a coal-fired power plant and a captive VCM plant to minimize the problems associated with the transportation of acetylene. The VCM plant would require a source of hydrogen chloride as well, which presumably will be available in the complex. An in-depth analysis of the cost of acetylene from coal would then require that all the costs involved in the associated plants be included. A simplifying assumption was made to decouple the acetylene and VCM plants from the rest of the complex. The capital costs assigned to the acetylene and VCM plants were based on the costs of the respective plants within battery limits only. Offsite capital costs were not considered. An equally good assumption would have set the offsite costs as a certain percentage of the associated capital costs. This would have resulted in increased product costs and led to an endless debate about the accuracy of the percentage figure chosen and the appropriateness of this figure to all the processes under consideration. Site-specific data and an analysis of the proposed chemical complexes would also have been required.

Working capital requirements for each of the production processes were aiso neglected. Estimates of working capital for each process are largely a function of feedstock and finished product inventories. A cost-minimizing operation would require site-specific trade-offs based on storage costs (including both working capital and capital investment in storage facility), reliability of supply, reliability of transportation and the potential increased productive capacity of the plant. By ignoring working capital, the actual costs will be lower. For ethanol production from sugar cane, however, the cost for ethanol was derived from Schaffer, where a six-month operating cycle was assumed. In the United States, where the sugar cane season is only three months long and economic considerations require year-round operation of the plant, the cost of ethanol is considerably higher than in warmer climates.

For the period of construction of the various plants, the CPI average of three years was utilized. During that period it was assumed that 20% of the required capital would be expended in the first year and 40% in each of the two succeeding years.

For plant life, the CPI average of 12 years was utilized for all processes. During the life of the plant it was assumed that the cost of the product would climb at the average yearly inflation rate of 6%. It was assumed that the construction of all plants would begin in 1979 and be in full production in 1982. Capital costs (per 1978) for each plant were obtained from the literature and made to escalate at 6% during construction. Capital costs were assumed to be independent of crude oil prices as the cost of feedstock is hypothesized to experience a step function change. The source of required capital was assumed to be 30.3% from debt (bank loans, bond issues, etc.) and the rest from equity (stocks), resulting in a debt:equity ratio of 43.5%, a reasonable estimate for the CPI. The cost of debt was estimated to be 10%, and the after-tax return on equity 18%, as is common for the CPI.

For tax purposes, a 50% corporate income tax rate was assumed and a property tax rate of 2.5% was utilized. The property tax rate was applied to the plant costs only, as offsite investments were defined to be zero. The usual investment tax credit of 10% was utilized. Table 6-1 lists all the financial parameters utilized for all the processes studied.

The cost of feedstock was incremented by constant steps. The cost of crude oil was used as the basis and was allowed to vary between $12/bbl, the 1978 price in the U.S., to $26/bbl in $2 increments. The costs of the other feedstocks, namely coal and sugar cane, were incremented by the same ratios. Table 6-2 shows the feedstocks costs utilized in this study.

ETHYLENE FROM NAPHTHA

Complete data are available for the ethylene from naphtha process and,

Table 6-1. Financial Parameters

Inflation Rate, %	6
Capital Components	
Debt, %	30.3
Cost of debt, %	10.0
Cost of equity, %	69.7
Taxes	
Income tax rate, %	50.0
Property tax rate, %	2.5
Investment tax credit, %	10.0
Plant Life, years	12.0
Tax Life, years	8.0

Table 6-2. Independent Variable Costs Utilized in the Study

Independent Variable	Normalized Cost of Input							
	1.0	1.17	1.3	1.5	1.67	1.83	2.0	2.17
Crude Oil, $/bbl	12.00	14.00	16.00	18.00	20.00	22.00	24.00	26.00
Sugar Cane, $/metric ton	11.00	12.83	14.67	16.50	18.33	20.17	22.00	23.83
Ethanol, $/$10^6$ Btu	14.16	15.67	17.19	18.70	20.20	21.72	23.23	24.74
Coal, $/short ton	28.00	32.67	37.33	42.00	46.67	51.33	56.00	60.67
Electricity[a], ¢/kWh	2.50	2.71	2.92	3.13	3.33	3.54	3.75	3.96

[a]Cost of electricity for the base case was assumed to be divided equally between cost of fuel, 1.25¢, and cost of production, 1.25¢. The cost of fuel was incremented only.

in this case, it was possible to estimate this process quite accurately. A 1 billion-lb/yr ethylene plant was hypothesized with an output of 960 million tons/yr of ethylene. The principal difficulty was that a number of valuable by-products are produced in addition to ethylene, and it was necessary to allocate credits to these by-products in estimating a cost* for ethylene. This difficulty was compounded because a naphtha cracking plant can be modified to change its product mix as dictated by market forces. For this study, it was assumed that the product mix remains constant for all costs of the input naphtha. The plant was then operated to produce ethylene, and a cost for this ethylene was determined. The by-product credits were then subtracted from the calculated ethylene cost to obtain the true cost of the ethylene. The 1978 prices of the by-products were utilized to determine the cost of ethylene produced from $12/bbl crude (Table 5-2). When the calculated cost of ethylene increased as a result of increased crude costs, the

---

*Cost includes the normal return in investment to equity owners. A market price in excess of cost would result in the economic phenomenon of excess profits.

by-product credits were increased by the same proportion as the increase in calculated ethylene price:

$$TC_E = CC_E - \frac{CC_E}{CC_{E\,1978}}\, C_{BP} \tag{1}$$

where TC_E =   true cost of ethylene
           CC_E =   calculated cost of ethylene
           $CC_{E\,1978}$ =   calculated cost of ethylene in 1978 when the cost of crude oil was \$12/bbl
           C_{BP} =   1978 cost of the by-products

This latter assumption is quite critical because it assumes that the total value of the by-products will not change with crude oil prices. However, within the context of this study, where crude oil is assumed to increase in step function fashion this assumption is valid.

Table 6-3 is a fact sheet describing all the assumptions for the ethylene from naphtha process. The naphtha costs were taken from Table 6-4. A description of the method used for computing the naphtha costs is included in Chapter 8.

Table 6-3. Ethylene from Naphtha Fact Sheet[a]

Input: Naphtha–55° API
Principal Product: Ethylene   Conversion Rate: 32.8% of naphtha input by weight

Principal By-products (Table 5-2)
Plant Capacity: 10^9 lb/yr of ethylene–Capacity Factor 96%
Cost of Plant Within Battery Limits: \$223 x 10^6
Offsite Costs: 0
Variable Costs: 0
Fixed Costs: \$48,894 x 10^6

---

[a]Overnight construction costs (1978 dollars) at a midcontinent U.S. location. Escalation and inflation during the construction cycle and an allowance for funds during construction must be added to derive the capital investment. The distinction between fixed and variable O & M costs is artificial and for convenience of using the existing MITRE Full Life Cycle Costing Model only. Capacity factors are always constant.

ETHYLENE FROM CANE SUGAR

Recently there has been renewed interest in the production of ethylene from sugar cane, especially in Brazil, where the production of ethanol to supplement gasoline in transportation applications is being actively pursued (Yang and Juchi, 1978). In the U.S., detailed studies have been conducted for the production of ethanol from biomass materials. Rather than duplicate these studied here, the cost of ethanol as a function of sugar

Table 6-4. Allocated Equilibrium Product Values as Dependent on Crude Price

Refinery Gate Arab. Light Crude Oil Price ($/bbl)	Refinery Gate Product Values, $/bbl		
	Naphtha	Pyrolysis Gasoline	Pyrolysis Fuel Oil
12	15.0	17.8	12.8
14	17.0	20.0	15.6
16	19.2	22.1	18.0
18	21.3	24.3	20.0
20	23.3	26.5	22.1
22	25.2	28.7	23.9
24	27.2	30.8	26.0
26	29.2	33.0	27.8

cane prices was obtained and utilized. A study by Schaffer showed that the cost of ethanol in $/million Btu consists of the following components: (1) capital costs, $2.73; (2) operating costs, $2.37; and (3) cost of sugar cane, $8.24, based on sugar cane at $10 per short ton. For our purposes, it is assumed that the capital and operating costs remain constant and the cost of sugar cane is incremented in step function fashion to determine the variation of the cost of ethanol. The base price (1978) utilized in this study is $11/metric ton, or $9.06/million Btu. The resulting cost of ethanol utilized is shown in Table 6-2 for the same incremented ratios utilized for crude oil.

The capital and operating costs of the ethanol dehydration plant were taken from Tsao and Reilly (1977) and Table 6-5 lists the data utilized in the computation of the price of ethylene. The operating costs include cost of

Table 6-5. Ethylene from Ethanol Fact Sheet[a]

Input: Ethanol–95° GL	
Principal Product: Ethylene	Conversion Rate: 95% of theoretical
Principal By-products: Water–no value	
Plant Capacity: 60,000 metric tons/yr	
Cost of Plant within Battery Limits: $12.75 x 10^6	
Offsite Costs: 0	
Variable Costs: 28.7414¢/lb	
Fixed Costs: 0	

[a]Overnight construction costs (1978 dollars) at a midcontinent U.S. location. Escalation and inflation during the construction cycle and an allowance for funds during construction must be added to derive the capital investment. The distinction between fixed and variable O & M costs is artificial and for convenience of using the existing MITRE Full Life Cycle Costing Model only. Capacity factors are always constant.

imported energy in the form of electricity and steam, which would not be applicable if bagasse were used. However, use of bagasse would require the construction of steam and electric generating plants.

ACETYLENE FROM COAL

As stated earlier, for the acetylene from coal process there are presently no operating plants that utilize the arc-coal process. The plant hypothesized for this study is a 300 million lb/yr acetylene plant costing $82.2 million. Every pound of acetylene requires 4.75 kWh of electricity, which means that approximately 200 MW of installed capacity would be required. The average size of a coal-fired plant is about 600 MW, and its efficiency with scrubbers is slightly more than 30%. Also, the average life of an electric power plant is 30 years, instead of the 12 years assumed for the acetylene plant. For this study, we considered including an electric generating plant as part of the complex; however, this presented the following difficulties:

1. The plant would have to have a nonstandard, 200-MW capacity and have a 12-year life, or it could be a 600-MW plant with a 30-year life and would produce electricity for sale outside the complex. For the first choice, the efficiency would have to be reduced, as lower priced, less efficient boilers would have to be specified. The cost of electricity would then have to be higher than the cost of electricity produced by the utility and, in actual costs to the complex, it may be higher than the cost of industrial electricity obtained from the utility. For the second choice, at the present capital cost of approximately $500/KW, the capital cost of the electric plant would be $300 million, or approximately four times the price of the acetylene plant. For the electricity exported offsite, some assumptions would have to be made about its disposition.

2. Another difficulty introduced by onsite generation is the relatively poor reliability (approximately 80%) of a coal-fired plant. An alternate supply of electricity would have to be provided to ensure full-time operation of the acetylene plant.

In view of these difficulties, it was assumed that electricity for the operation of acetylene plant would be purchased from a utility and was introduced in the calculations as a variable cost. The 1978 price of electricity was taken to be 2.5¢/kWh. This cost includes the cost of fuel to produce the electricity, among others. The cost of acetylene was calculated as a function of the price of electricity with the portion of the cost attributed to fuel increasing at the same proportion as the crude oil price. The portion of the cost for fuel was assumed to be 1.25¢ (50%). The escalated values are shown in Table 6-2.

As for the coal, the price chosen was that of low sulfur coal at $28 per ton, which at the time of this writing is the price of delivered coal in the U.S. It should be noted, however, that if the acetylene plant is located at the mine mouth, considerably less expensive coal would be available. Fortunately the sensitivity of the acetylene cost to the price of coal was found to be quite low, as will be shown later. In addition to calculating the cost of acetylene with the price of coal and the price of electricity increasing individually, runs were made in which both were escalated simultaneously.

In the production of acetylene, by-products are produced, the most important of which are carbon black, hydrogen cyanide (HCN), char and fuel gas. These were treated similarly to the by-products of the naphtha cracking plant. Table 6-6 lists the information utilized in the computation of acetylene cost.

Table 6-6. Acetylene from Coal Fact Sheet[a]

Input: Coal—25 million Btu/ton
Principal Product—Ethylene   Conversion Rate: 30% of coal input by weight
Principal By-products: Carbon black, hydrogen cyanide, char and fuel gas
Plant Capacity: 300×10^6 lb/yr of acetylene—Capacity Factor 100%
Cost of Plant within Battery Limits: 82.2×10^6
Offsite Costs: 0
Variable Costs: Electricity—4.75 kWh/lb of acetylene—11.876¢ lb
FixedCosts: 24.118×10^6

[a]Overnight construction costs (1978 dollars) at a midcontinent U.S. location. Escalation and inflation during the construction cycle and an allowance for funds during construction must be added to derive the capital investment. The distinction between fixed and variable O & M costs is artificial and for convenience of using the existing MITRE Full Life Cycle Costing Model only. Capacity factors are always constant.

VCM FROM ACETYLENE

To produce VCM from acetylene, hydrogen chloride is required. Normally, hydrogen chloride is available in a large chemical complex, and could be utilized with little effect on product cost. Hence, for the purpose of this chapter it was assumed that hydrogen chloride would be available at no cost. Table 6-7 lists the input data for the VCM plant and Figure 6-2 shows the material balance. The VCM plant was assumed to have a conversion rate of 99% of theoretical.

VCM FROM ETHYLENE

Both chlorine and oxygen are required in the production of VCM from ethylene. The demand and cost of chlorine have been increasing because of increased demand for PVC, and as long as the latter is to be produced from VCM manufactured from ethylene, chlorine will be in demand.

Chlorine is manufactured by electrolytic decomposition of sodium chloride, so is greatly dependent on the cost of electricity. However, for the

Table 6-7. VCM from Acetylene Fact Sheet[a]

Input: Acetylene
           Hydrogen Chloride—no cost
Principal Product: VCM   Conversion Rate: 99% of theoretical
                                          by weight
By-products: None
Plant Capacity: 717 x 10^6 lb/yr of VCM—Capacity Factor 100%
Cost of Plant within Battery Limits: $21.7 x 10^6
Offsite Costs: 0
Variable Costs: 0.023¢/lb
Fixed Costs: $42.20 x 10^6

[a]Overnight construction costs (1978 dollars) at a midcontinent U.S. location. Escalation and inflation during the construction cycle and an allowance for funds during construction must be added to derive the capital investment. The distinction between fixed and variable O & M costs is artificial and for convenience of using the existing MITRE Full Life Cycle Costing Model only. Capacity factors are always constant.

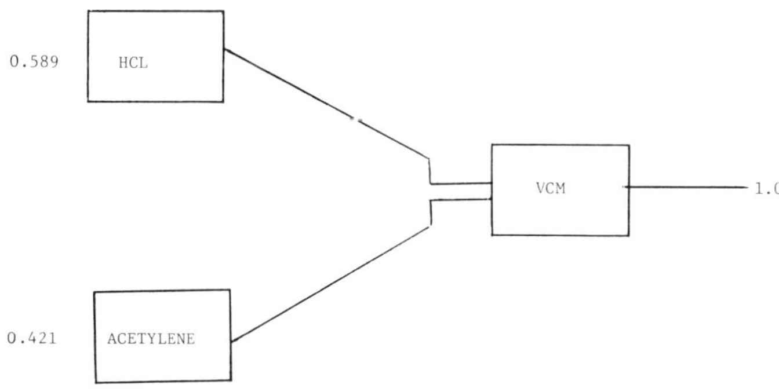

Figure 6-2. Material balance for the manufacture of VCM from acetylene.

purpose of this chapter, where crude oil price is assumed to experience a step function increase it is reasonable to assume that the cost of chlorine would remain constant.

Table 6-8 lists the input data utilized in the calculation of the cost of VCM from ethylene and Figure 6-3 shows the material balance. The VCM plant was assumed to have a 95% conversion rate. The principal by-product is water, which was assumed to have no value.

Table 6-8. VCM from Ethylene Fact Sheet[a]

Inputs: Ethylene, Chlorine, Oxygen
Principal Product: VCM   Conversion Rate: 95% of theoretical
                                          by weight
Principal By-Products: Water—no value
Plant Capacity: 714 x 10^6 lb of VCM/yr—Capacity Factor 100%
Cost of Plant within Battery Limits: \$42.2 x 10^6
Offsite Costs: 0
Variable Costs: Chlorine 6.8¢/lb, oxygen 1.2¢/lb-4.4612¢/lb of VCM
Fixed Costs: \$15.3645 x 10^6

---

[a]Overnight construction costs (1978 dollars) at a midcontinent U.S. location. Escalation and inflation during the construction cycle and an allowance for funds during construction must be added to derive the capital investment. The distinction between fixed and variable O & M costs is artificial and for convenience of using the existing MITRE Full Life Cycle Costing Model only. Capacity factors are always constant.

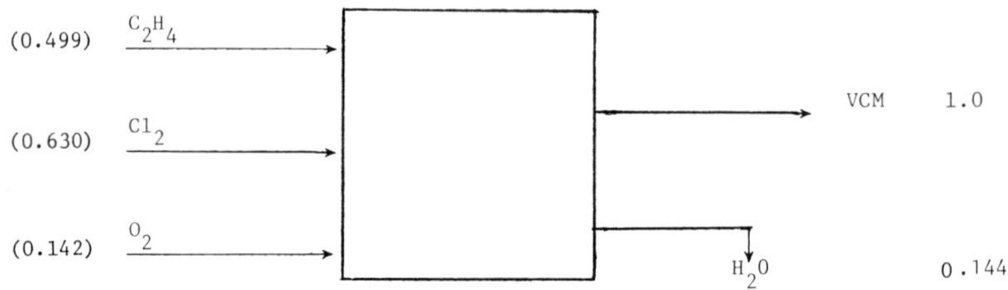

Figure 6-3. Material balance for the production of VCM from acetylene.

PREPARATION AND PRESENTATION OF RESULTS

Using the computer model described in detail in Chapter 9, the cost of the various products was calculated based on the data listed in Tables 6-1 to 6-8. With the hypothesized 1979 construction date, the 1982 product costs were calculated; 1978 costs were obtained by deflating the 1982 prices by 6%/yr.

The computer program can calculate product costs for either of the following assumptions:

1. The product cost inflates at the annual rate of 6% during the 12-year life of the plant.
2. The product cost remains constant over the life of the plant.

Obviously, the product costs calculated using the first assumption will be the smaller of the two. Table 6-9 displays this difference. Case 1 shows the product costs using the first assumption. The by-product credits were not subtracted from the costs of ethylene from naphtha and acetylene from coal. The acetylene from coal price is that for variable coal and electricity costs. The difference in cost between the two assumptions is considerable, ranging between 7.7¢/lb and 12.4¢/lb for ethylene and naphtha, and 9.7¢/lb and 16.2¢/lb for ethylene from ethanol. This points out the importance of the assumptions that must be made on the behavior of the cost of the product during the life of the plant. It is believed that the costs calculated using the first assumption are more realistic because they simulate a real-world situation, so they are used here. A 6% inflation rate is reasonable considering the historical behavior of inflation.

Table 6-9. Calculated Product Costs (1978, ¢/lb)

Normalized Cost of Input	Ethylene From Crude[a]		Ethylene From Ethanol		Acetylele From Coal Case C[a]	
	Case 1[b]	Case 2[c]	Case 1[b]	Case 2[c]	Case 1[b]	Case 2[c]
1.0	26.95	34.65	34.57	44.45	30.07	38.66
1.17	29.48	37.90	37.73	48.51	31.86	40.96
1.3	31.77	40.84	40.90	52.59	33.62	43.22
1.5	34.19	43.94	44.01	56.64	35.40	45.51
1.67	36.48	46.89	47.19	60.67	37.13	47.72
1.83	38.65	49.69	50.37	64.75	38.90	50.01
2.0	40.95	52.64	53.52	68.81	40.67	52.28
2.17	43.25	55.60	56.67	72.86	42.45	54.57

[a]No by-product credits. Variable coal and electricity costs.
[b]Cost of product allowed to inflate at 6% over the life of the plant.
[c]Cost of product constant over life of the plant.

Tables 6-10 and 6-11 show the cost of ethylene and acetylene as a function of the variation in costs of the input. The exclusion of offsite and operating capital costs resulted in unrealistically low cost for the products. The inclusion of these costs could add 3–5% to the cost of the product, depending on the location of the site. In Tables 6-12 to 6-14, the contributions to the product costs of the four components—feedstock, operation and maintenance, capital costs and taxes—are shown for the case where the product cost is allowed to inflate over the life of the plant. In the case of ethylene (Table 6-12), the principal contribution is from the cost of feedstock, with the case of ethanol showing the highest value. In the case of acetylene from coal, the largest contribution is from O & M costs. This is not surprising as the cost of electricity to manufacture the acetylene is included in this cost. This is particularly obvious in the O & M costs of Case B, where the price of electriciy was escalated. For Case C, where both the cost of electricity and that of coal were escalated, the feedstock and O & M costs increased, but not at the same rate as when the coal price and the electricity costs were increased independently. That a large portion of the costs are due to O & M leads to the conclusion that improvements in technology could significantly decrease the cost of acetylene, while in cases where the largest portion of the cost is due to the feedstock (e.g., with ethylene), technology improvements would have a less significant effect.

Table 6-15 shows the cost distribution in VCM manufacture. Here again, feedstock costs are important. In the VCM from ethylene case the O & M costs are higher, due partly to the fact that they include the cost of chlorine and oxygen. Although the cost of chlorine was assumed to remain constant with increased costs of feedstock, the expected increased demand for this chemical will certainly result in its increased cost.

Table 6-10. Cost of Ethylene Produced from Naphtha and from Ethanol (1978)

| Normalized Cost of Input | Product Costs (¢/lb) | | | |
| | Ethylene/Naphtha | | | Ethylene/ Sugar Cane Cost |
	Calculated Cost	Credits	Cost	
1.0	26.95	15.70	11.25	34.57
1.17	29.48	17.17	12.31	37.73
1.3	31.77	18.50	13.27	40.90
1.5	34.19	19.91	14.28	44.01
1.67	36.48	21.24	15.24	47.19
1.83	38.65	22.51	16.14	50.37
2.0	40.95	23.85	17.10	53.52
2.17	43.25	25.19	18.06	56.67

The costs of VCM manufactured from ethylene and acetylene were calculated and are shown in Tables 6-10 and 6-11, respectively; both are shown in Table 6-14. These data are plotted in Figures 6-4 and 6-5. A straight line was fitted to the data for the cost of VCM as a function of the cost of ethylene, and the equivalent cost of ethylene for acetylene was computed using this linear equation. The relationship between the cost of VCM and that of ethylene was found to follow the following equation:

$$C_{VCM} = 0.5\, C_E + 8.59 \tag{2}$$

where C_{VCM} and C_E are the costs of VCM and ethylene, respectively. Table 6-16 summarizes the results of the study; the costs of ethylene for escalating

Table 6-11. Variations in the Cost of Acetylene as a Function of Variations in the Cost of Coal and Electricity (1978, in ¢/lb)

Normalized Cost of Input	Case A[c]			Case B[b]			Case C[c]		
	Calculated Cost	Credits	Cost	Calculated Cost	Credits	Cost	Calculated Cost	Credits	Cost
1.0	30.07	8.30	21.77	30.07	8.30	21.77	30.07	8.30	21.77
1.17	30.85	8.52	22.33	31.08	8.58	22.50	31.86	8.79	23.07
1.3	31.63	8.73	22.90	32.06	8.85	23.21	33.62	9.28	24.34
1.5	32.40	8.94	23.46	33.07	9.13	23.94	35.40	9.77	25.63
1.67	33.18	9.15	24.06	34.01	9.39	24.62	37.13	10.25	26.88
1.83	33.96	9.36	24.60	35.02	9.67	25.35	38.90	10.74	28.16
2.0	34.73	9.57	25.16	36.01	9.94	26.07	40.67	11.23	29.44
2.17	35.51	9.78	25.73	37.01	10.22	26.79	42.45	11.72	30.73

[a]Coal cost variable only.
[b]Electricity cost variable only.
[c]Coal and electricity costs variable.

Table 6-12. Percent Cost Distribution in the Production of Ethylene from Naphtha and from Ethanol

Normalized Cost of Input	Ethylene/Naphtha				Ethylene/Ethanol			
	Feed	O & M	Capital	Taxes	Feed	O & M	Capital	Taxes
1.0	63.83	18.89	14.06	3.22	85.56	9.31	4.98	1.14
1.17	66.92	17.28	12.86	2.94	86.77	7.62	4.57	1.05
1.3	69.31	16.03	11.93	2.74	87.80	7.03	4.21	0.96
1.5	71.47	14.90	11.09	2.54	88.67	6.52	3.91	0.89
1.67	73.27	13.96	1.039	2.38	89.42	6.09	3.65	0.84
1.83	74.77	13.18	9.81	2.24	90.09	5.71	3.42	0.78
2.0	76.19	12.44	9.26	2.12	90.67	5.37	3.22	0.74
2.17	77.45	11.78	8.77	2.01	91.19	5.07	3.04	0.70

Table 6-13. Percent Cost Distribution in the Production of Acetylene from Coal

Normalized Cost of Input	Case A[a]				Case B[b]				Case C[c]			
	Feed-stock	O & M	Capital	Taxes	Feed-stock	O & M	Capital	Taxes	Feed-stock	O & M	Capital	Taxes
1.0	15.50	66.20	14.87	3.40	15.50	66.22	14.87	3.40	15.50	66.22	14.87	3.40
1.17	17.63	64.55	14.49	3.32	15.00	67.32	14.39	3.29	17.07	65.68	14.03	3.21
1.3	19.65	62.97	14.14	3.24	14.54	68.33	13.94	3.19	18.49	65.17	13.30	3.04
1.5	21.58	61.46	13.80	3.16	14.10	69.28	13.52	3.09	19.76	64.72	12.63	2.89
1.67	23.43	60.02	13.48	3.08	13.71	70.14	13.15	3.01	20.93	64.27	12.04	2.76
1.83	25.17	58.65	13.17	3.01	13.31	70.99	12.77	2.92	21.97	63.90	11.49	2.63
2.0	26.84	57.34	12.87	2.95	12.95	7.179	12.42	2.84	22.93	63.56	10.99	2.52
2.17	28.45	56.08	12.59	2.88	12.60	72.55	12.08	2.77	23.80	63.26	10.53	2.41

[a]Coal cost variable only.
[b]Electricity cost variable only.
[c]Coal and Electricity costs variable.

Table 6-14. Percent Cost Distribution in the Production of VCM from Acetylene and from Ethylene

Normalized Cost of Input	VCM/Acetylene				VCM/Ethylene			
	Feed-stock	O & M	Capital	Taxes	Feed-stock	O & M	Capital	Taxes
1.0	58.34	37.71	3.21	0.73	39.50	49.41	9.22	1.87
1.17	—	—	—	—	41.67	47.64	8.89	1.81
1.3	58.96	36.60	3.11	0.71	43.51	46.14	8.61	1.75
1.5	60.15	36.08	3.07	0.70	45.32	44.66	8.33	1.69
1.67	60.75	35.53	3.02	0.69	46.93	43.34	8.09	1.64
1.83	61.28	35.05	2.98	0.68	48.37	42.17	7.87	1.60
2.0	61.81	34.57	2.94	0.67	49.81	40.99	7.65	1.55
2.17	62.34	34.09	2.90	0.66	51.17	39.87	7.44	1.51

Table 6-15. Cost of VCM (1978, in ¢/lb)

Normalized Cost of Input	Feedstock			
	Ethylene	Acetylene Case A[a]	Acetylene Case B[b]	Acetylene Case C[c]
1.0	14.21	15.71	15.71	15.71
1.17	14.74	15.94	16.02	16.25
1.3	15.22	16.18	16.32	16.79
1.5	15.72	16.42	16.63	17.33
1.67	16.21	16.67	16.90	17.87
1.83	16.65	16.90	17.22	18.40
2.0	17.13	17.13	17.52	18.94
2.17	17.61	17.38	17.82	19.47

[a]Coal cost variable only.
[b]Electricity cost variable only.
[c]Coal and electricity costs variable.

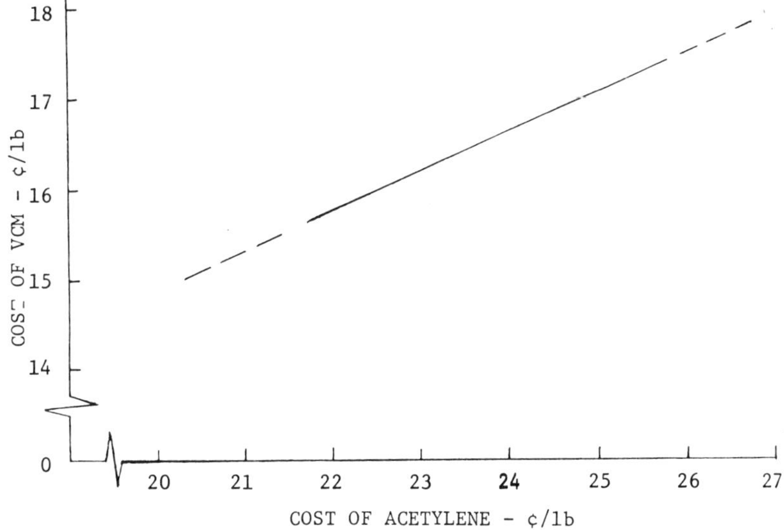

Figure 6-4. Cost of VCM as a function of the cost of acetylene.

costs of the input are shown. The normalized input cost, 1.0, represents 1978 conditions:

Crude Oil @	$12.00/bbl
Sugar Cane @	$11.00/metric ton
Coal @	$28.00/short ton
Electricity @	2.5¢/kWh

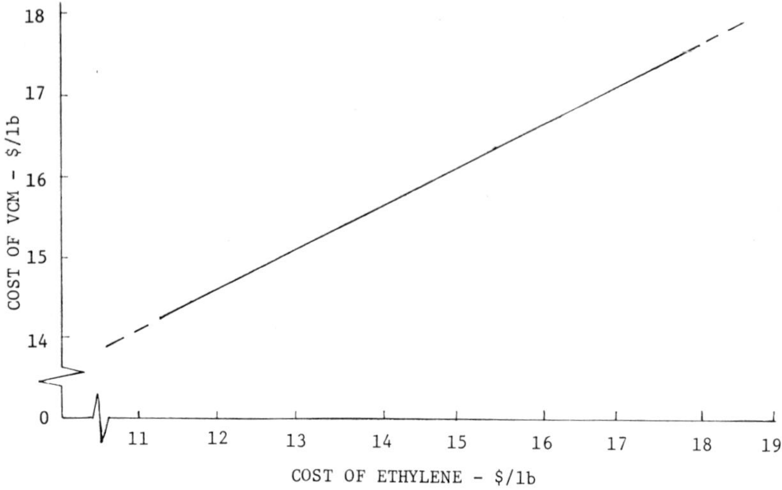

Figure 6-5. Cost of VCM as a function of the cost of ethylene.

Table 6-16. Cost of Ethylene (¢/lb)

Normalized Cost of Input	Ethylene/ Naphtha	Ethylene/ Ethanol	Ethylene/ Coal Case A[a]	Ethylene/ Coal Case B[b]	Ethylene/ Coal Case C[c]
1.0	11.25	34.57	14.25	14.25	14.25
1.17	12.31	37.73	14.71	14.87	15.33
1.3	13.27	40.90	15.19	15.47	16.42
1.5	14.28	44.01	15.68	16.10	17.50
1.67	15.24	47.19	16.18	16.64	18.58
1.83	16.14	50.37	16.64	17.28	19.64
2.0	17.10	53.52	17.10	17.88	20.72
2.17	18.06	56.67	17.60	18.48	21.78

The succeeding values of the input are given in Table 6-2. These data were plotted in Figure 6-6 to display the cost relationship. Each line in Figure 6-6 should be read individually, e.g., if the cost of crude doubles and the cost of coal, sugar cane and electricity remain the same (as in 1978), the cost of ethylene will be 17.10¢/lb, 34.52¢/lb and 14.25¢/lb for ethylene made from crude oil, sugar cane and coal, respectively.

The costs for ethylene shown in Figure 6-6 were fitted with straight lines. The equations of these lines follow:

$$C_{eo} = 5.84\, C_o/12 + 5.46 \tag{3}$$

$$C_{es} = 18.94\, C_s/11 + 15.63 \tag{4}$$

$$C_{ec} = 2.87\, C_c/28 + 11.37 \tag{5}$$

$$C_{el} = 3.62\, (C_l - 1.25)/1.25 + 10.63 \tag{6}$$

$$C_{ecl} = 6.45\, C_c/28 + 7.81 \tag{7}$$

where
C_{eo} = the cost of ethylene from crude oil, $\cent$/lb
C_o = the price of crude oil, \$/bbl
C_{es} = the cost of ethylene sugar cane, $\cent$/lb
C_s = the price of sugar cane, \$/metric ton
C_{ec} = the equivalent cost of ethylene from coal with the price of coal variable
C_c = the price of coal, \$/short ton
C_{el} = the equivalent cost of ethylene from coal with the price of electricity variable
C_l = the price of electricity, $\cent$/kWh
C_{ecl} = the equivalent cost of ethylene with coal and electricity price variable

Equation 7 actually can be derived from Equations 4 and 5 as follows:

$$C_{ecl} = C_{ec} + C_{el} - 14.25 \tag{8}$$

The constant, 14.25, in Equation 8 is the base cost of ethylene in 1978. Substituting Equations 5 and 6 in 8 yields:

$$C_{ecl} = 6.49\, C_c/28 + 7.85 \tag{9}$$

The errors in the constants in Equation 9 when compared to Equation 7 are less than 0.8% and can be attributed to round-off errors.

DISCUSSION

To determine the sources of errors, we begin by listing the assumptions on which the study was based:

1. The cost of feedstock for the different processes will increase independently of other costs.
2. The product mix of each process is independent of feedstock costs.
3. The values of the by-products escalates at the same rate as the cost of the product.
4. Offsite and working capital costs are neglected.

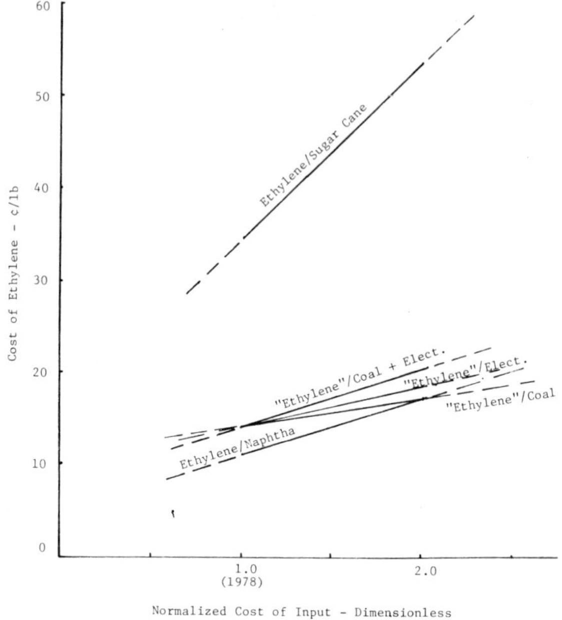

Figure 6-6. Cost of ethylene produced by various processes.

5. The cost of the product inflates over the life of the plant, which is 12 years.

6. It is possible for the costs of the coal and electricity to escalate at different rates.

7. Representative financial parameters for the CPI in the United States have been used.

8. The base costs (1978) for the inputs to the various processes were $12/bbl, $11/metric ton, $28/short ton, and 2.5¢/kWh for Arabian light crude, sugar cane, coal and electricity, respectively.

Assumption 1 will have the least effect on those products where the contribution of feedstock to the cost is largest. Tables 6-12 to 6-14 show the cost contribution of the various cost elements to the cost of the product. Ethylene from ethanol begins with a feedstock cost contribution of more than 85%. In this case, variations in the other costs will have little effect on the cost of the product, e.g., an increase of 10% in the capital costs will have a 0.5% effect on the cost of the product. Similarly, in the case of ethylene from naphtha, the feedstock contribution is approximately 64%, which means that a 10% increase in other costs will affect the cost of ethylene by approximately 3.6%. On the other hand, in the case of acetylene from coal, the contribution to the cost of acetylene from the price of coal is approximately 15%, and from electricity 40% (4.75 kWh/lb of acetylene are required at 2.5¢/kWh; the contribution of electricity to the cost of acetylene is approximately 12¢, or approximately 40% of the calculated cost of acetylene.) This leaves 45% for the remaining cost elements. In this case, a 10% increase in these costs would result in a 4.5% cost increase. However, production of acetylene in the arc process is not commercial, and the uncertainty in the engineering cost estimates utilized in this study would probably overshadow any possible escalation resulting from increased feedstock costs. In the case of VCM (Table 6-14), the feedstock costs are approximately 50% of the VCM cost for the base case. VCM technology is well established and, in this case, increased costs of capital and O & M that could result from increased feedstock costs could significantly affect the cost of the product. The O & M contribution to the cost of VCM from ethylene includes the cost of chlorine. The contribution of the chlorine to the cost of VCM is approximately 4¢/lb for chlorine at 6¢/lb, or approximately 26%. An increase in the cost of chlorine, which is imminent because of increased demand, coupled with an increase in O & M costs, which would result from increased feedstock costs, could significantly increase the cost of VCM from ethylene and distrupt its relationship to the cost of VCM from acetylene. The net effect would be to improve the economics of an acetylene-based carbochemical industry. However, under normal conditions, chlorine would be generated within the complex and should cost less than the assumed 6¢/lb.

Assumptions 2 and 3, that the product mix and by-product costs are independent of feedstock costs, will affect the cost of ethylene from crude oil and that of acetylene from coal. The process of production of ethylene from sugar cane has no by-products, although Yang et al. (1978) proposed that furfural and bagasse could possibly be produced in an ethylene from sugar cane complex. The product mix and by-product cost allocation assumption should affect the cost of ethylene more than the cost of acetylene because the ratio of the cost of by-products to cost of product for ethylene is larger than for acetylene (1.4 and 0.4, respectively). However, in the case of ethylene from oil there are many by-products, and although the cost of one of those by-products may increase because of market forces, the total cost of the by-products will increase to a lesser extent. On the other hand, the cost of the by-products of acetylene is based on the assumption that the technology for separating carbon black from char will be developed. If not, the char-carbon black mixture will have to be sold at char prices, increasing the cost of acetylene by 2–3¢.

Assumption 4, that offsite and operating capital costs are neglected, resulted in lower costs for all the products. It does not necessarily reduce the cost of all products by the same amount or ratio. Depending on location of the site and on the desired balance between supplies and inventories that would take into consideration outside conditions, such as reliability of supply, efficacy of the transportation system, etc., these costs will vary between plants producing the same product or variety of products considered in this study.

Assumption 5, that the cost of the product will inflate at 6% during the life of plant, is believed to be realistic and simulates the real-world environment.

It is believed that assumption 6 (costs of coal and electricity could escalate at different rates in the production of acetylene) is also valid. Since electricity could be produced from a number of "feedstocks," it is possible that the cost of electricity would not necessarily vary at the same rate as that of coal. However, to account for the possibility that the acetylene plant could be located in a carbochemical complex that included a coal-fired electric plant, data were calculated for the case where one-half the cost of electricity would escalate at the same rate as the cost of coal; the remaining half would remain constant, as it was attributed to costs other than fuel costs.

Assumption 7 is also valid within the scope of this study.

CHAPTER 7

IMPACT OF TECHNOLOGY IMPROVEMENT

INTRODUCTION

All technology improves over time, even if only incrementally. Economy, efficiency and environmental considerations are the major driving forces for technology improvement in the U.S. In the absence of a major technological breakthough, improvements are more rapid at the birth of a technology and less significant with a mature technology.

We can foresee technological improvements in the three processes under consideration.

ETHYLENE FROM CRUDE OIL

With the increased dependence of the petrochemical industry on the refining industry for the naphtha and gas oil feedstocks, it is not surprising that processes offering less dependence are being investigated. As mentioned before, Union Carbide, collaborating with Kureha Chemical Industry Co. and Chiyoda Chemical Engineering and Construction Co. of Japan, has developed an advanced cracking reactor technology (ACR), primarily for ethylene production. A selected crude oil or distillate is injected into a chamber with high-temperature combustion gases. The vaporized feedstock and combustion gases pass through a venturi reaction chamber, where adiabatic cracking occurs. The reaction products are quenched rapidly and sent along for further processing (*Chem. Eng. News*, 1978).

The technology produces 60–70% yields of high-value chemical products, including more than 30% ethylene. The largest test so far has been a

100 million lb/yr ethylene reactor. To further minimize technical risks and to complete the development program, Union Carbide has constructed a $15 million ACR prototype unit at Seadrift, Texas, primarily to test long-term equipment operability. Capital investments are expected to be slightly higher than for conventional cracking plants. Dow (1977) has also announced the development of a process involving partial oxidation of the crude oil with oxygen and steam at similar reaction conditions.

Both developments anticipate a 1985 commercialization. Early economic estimates suggest a 1–3¢/lb ethylene manufacturing cost advantage of the whole crude process relative to naphtha cracking (*Chem. Wk.*, May 11, 1977).

One Conoco process for making ethylene from higher hydrocarbon cracking is being tested in a pilot plant at Naphtachimie's chemical complex near Marseilles. There are four German ethylene-from-crude processes, developed by BASF, Hoechst, Lurgi and Koppers about 15 years ago. All four technologies were successful on a pilot-plant scale, but were never commercialized for economic reasons.

ETHYLENE FROM BIOMASS

MITRE's study (Park et al., 1978) on alcohol production from biomass indicates that some 70% of the cost is associated with the raw material. Technological improvements are possible. The Brazilians estimate a 20–50% improvement in cost of ethylene with improved technology and expanded distillery operation (Yang and Juchi, 1978).

The C. E. Lummus Company has developed a fluidized bed system for the dehydration of ethanol to ethylene. Efficient temperature control, heat and mass transfer and continuity of operations are some of the advantages. A pilot-plant operation indicated a 99% yield by comparison with 94–96% for fixed bed systems. A substantially lower investment cost is also claimed—$6.1 million vs $17.750 million for a 60,000 metric ton/yr plant (Tsao and Reilly, 1977).

ETHYLENE/ACETYLENE FROM COAL

A large variety of chemicals can be produced directly from coal or indirectly from the synthesis gas derived from coal gasification. There are several possible processes for producing ethylene from coal:

- Cracking of coal liquids
- Variations of the Fisher-Tropsch Reaction Schemes
- Methanol homologation to ethanol followed by dehydration
- Cracking dimethyl ether
- Direct synthesis from a carbon monoxide-hydrogen mixture

Acetylene from crude oil and natural gas is also possible:

- Crude oil ACR (Union Carbide)—acetylene and ethylene coproducts.
- Naphtha two-stage cracking—acetylene and ethylene coproducts.
- Natural gas thermal cracking.

The economics of these processes do not look good at this time, however (*Chem. Eng. News*, November 21, 1977; Guccione, 1977).

As for the production of acetylene from coal, much improvement in the technology is possible. The basic AVCO process and its predecessors have been described extensively in a MITRE study (Barbier, 1976). Since then, the U.S. Department of Energy (DOE) has awarded $5 million to AVCO to rescucitate its process.

AVCO believes that many improvements in the reactor design and construction are feasible. It is also believed that the "Balance-of-Plant" can be substantially improved, including coal griding, coal feeding, adjustable variable pressure and quench rates.

The Orbach Reactor (U.S. Patent 3,332,870) appears to give higher acetylene yield (34.5% by weight acetylene and 12.0% ethylene) than the AVCO process. Plasma Chem Company has also several patents for improvements in reactor design for synthetization of gases (U.S. Patent 3,954,954; 3,840,750).

The concept of a coal refinery is making progress. Chemicals as well as fuel from coal gasification and liquefaction will have to be a reality by the late 1990s.

CHAPTER 8

NAPHTHA FEEDSTOCK PRICE PROJECTION

INTRODUCTION

To relate the cost of ethylene manufacture to the price of crude oil on the U.S. Gulf Coast, it is necessary to define the future marginal source of ethylene supply. Due to the declining availability of gas liquids in the United States in general and the Gulf Coast in particular, all but one of the plants under construction are designed for naphtha and/or gas oil feedstock. Hence, marginal economics should be based on an ethylene plant capable of processing either naphtha or gas oil. Since Middle Eastern crudes are in plentiful supply and since the naphtha from such crudes have superior feedstock quality for ethylene cracking, Arabian light naphtha was chosen as the feedstock for this marginal ethylene plant.

This naphtha is derived from refining Arabian light crude oil, thus producing a large number of other products. A basis for allocating the refining costs to the naphtha product is required. Furthermore, as shown in Table 8-1, the values for numerous coproducts are important in establishing the value of ethylene. Two of these products, pyrolysis gasoline and fuel oil, are often absorbed into the refinery operations. Because of the high octane characteristics of pyrolysis gasoline, it is commonly valued equivalent to unleaded gasoline. Hence, the values of refinery-produced unleaded gasoline and intermediate sulfur content fuel oil must also be established, along with the naphtha value.

THEORY OF REFINERY COST ALLOCATION

The United States petroleum industry, including certain offshore Caribbean refineries serving east coast import requirements, is composed of four

Table 8-1. Olefin Plant Yields Methane:Ethylene Naphtha Feedstock

Product	Yield (wt %)
Hydrogen	1.2
Methane	16.8
Ethylene	32.8
Propylene	12.7
Other C_3	0.8
Butadiene	5.1
Other C_4	4.0
Pyrolysis Gasoline[a]	21.4
Fuel Oil	5.2
	100.0

[a]Including BTX.

basic refinery types. The Topping Refinery (Figure 8-1) consists principally of a distillation column and associated offsites to separate crude oil directly into the primary products of naphtha, gas oil and fuel oil. If the crude oil is of high sulfur content, these products will be of similarly high sulfur content. Such refineries are used for production of specialty products, such as naphtha (for petrochemical feedstocks) or asphalt.

If gasoline production is required, the addition of a catalytic naphtha reformer is common, thereby creating a Hydroskimming Refinery (Figure 8-2). If a low-sulfur crude is used, such a refinery would produce gasoline and low-sulfur fuel oil products. With a high-sulfur crude, either high-sulfur fuel oils or specialty products such as asphalt are produced.

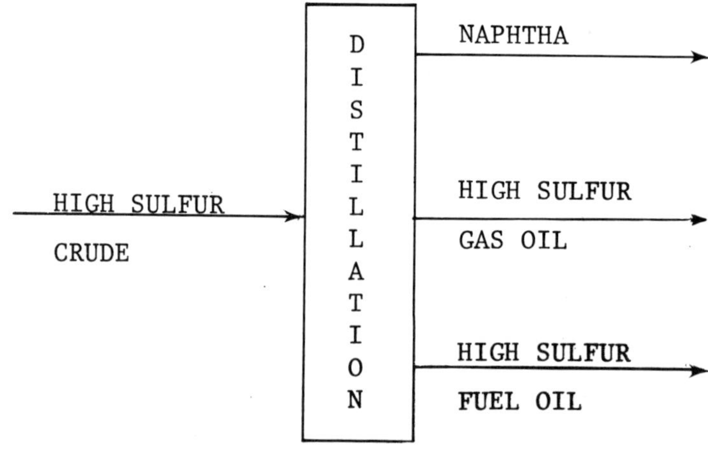

Figure 8-1. Topping refinery schematic.

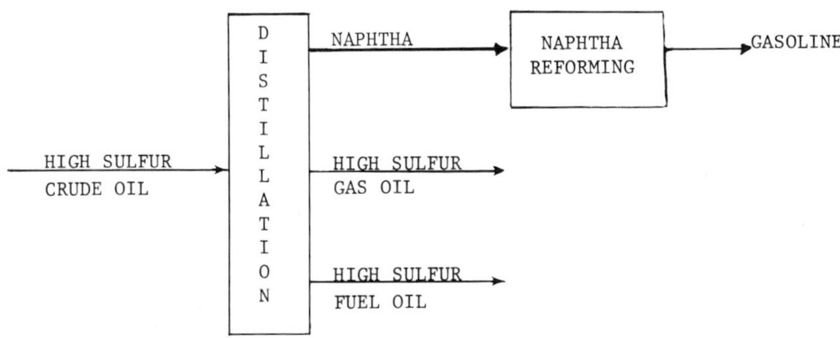

Figure 8-2. Hydroskimming refinery schematic.

With the current environmental restrictions on the use of high-sulfur fuel oils, it is becoming increasingly common in the United States and Caribbean refineries to desulfurize the fuel oils produced from high-sulfur crude oils. Since high-sulfur crude oil (such as Arabian light) will be used increasingly in the U.S. refineries, it is not surprising that two major U.S. refinery expansions have employed fuel oil desulfurization, converting them to fuels refineries similar to that shown in Figure 8-3.

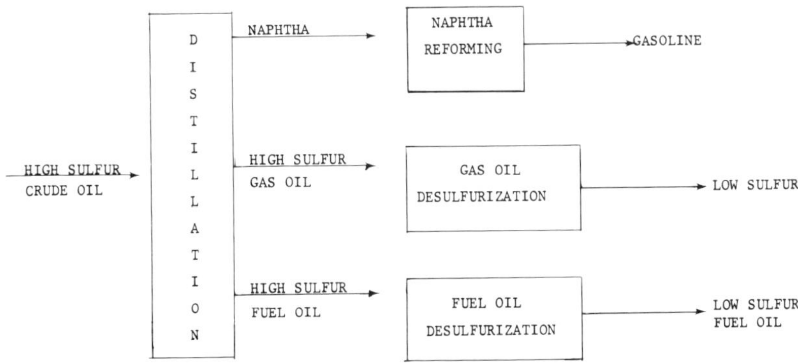

Figure 8-3. Fuels refinery schematic.

In the United States, the demand for gasoline has substantially exceeded the supply available naturally from crude oil through refinery processing types typified in Figures 8-2 and 8-3. Conversion refineries (Figure 8-4) have thus been constructed to convert excess fuel oil into gasoline by catalytic cracking or hydrocracking, providing a historic average gasoline yield of 50% of crude processed.

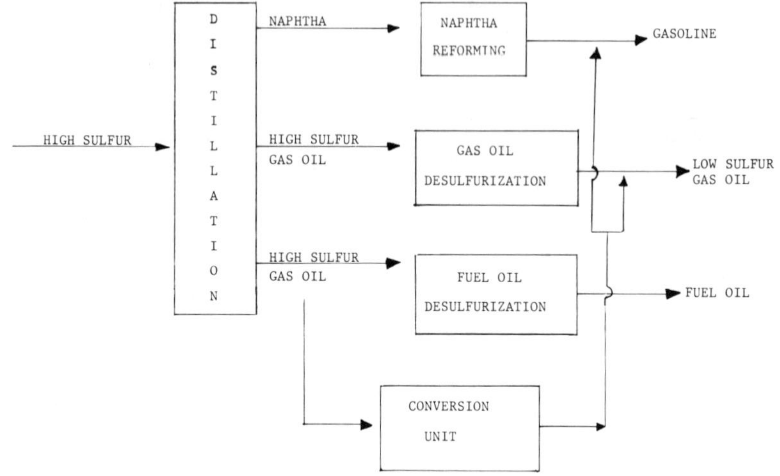

Figure 8-4. Conversion refinery schematic.

To make projections of product prices, econometric models of the composite of these refinery types representing U.S. industry must be developed, reflecting both the anticipated changes in crude oil distillation and other processing costs, as well as the future profitability of refining operations. For example, Table 8-2 illustrates the effect of refinery profitability on required product prices for one of these refinery types—a conversion refinery processing Arabian light crude oil. Refinery profitability is measured by a capital charge factor, reflecting annual cash flow necessary to:

1. repay debt financing, including interest;
2. pay taxes, after allowance for depreciation and interest; and
3. earn a return on equity in the project.

Under conventional fiscal regimes with about 50% corporate tax rates, a capital charge of around 25% is necessary to yield a 12–15% discounted cash flowrate of return over a 20-year project life.

The conversion refinery, optimized by means of a linear programming algorithm, will produce about 93,000 bbl/day of products from 100,000 bbl/day of crude (the remaining 7000 bbl/day being consumed in refinery fuel and losses), consuming no natural gas, purchased natural gasoline or butanes. A zero percent capital charge is reflected in the set of product prices, which average $13.72/bbl, yielding $1.276 million/day of total revenues. At a crude oil cost of $12/bbl, fixed operating costs of 51¢/bbl

Table 8-2. Illustrative Cash Flow Summary for Conversion Refinery
(100,000 bbl/SD Arabian light crude oil, $446 million investment)

	Capital Charge on Total Refinery bbl/day	0%		8%		20%	
		$/bbl/day	$1000/ day	$/bbl/day	$1000/ day	$/bbl/day	$1000/ day
Products							
Gasoline	46.7	14.27	666	15.40	719	17.10	799
Low-Sulfur Gas Oil	29.8	13.52	403	14.71	438	16.50	492
0.5% Sulfur Fuel Oil	16.5	12.52	207	13.71	226	15.50	256
TOTAL	93.0	13.72	1276	14.88	1384	16.62	1546
Costs							
Crude Oil Cost	100.0	12.00	1200	12.00	1200	12.00	1200
Fixed Operating Costs	–	0.51	51	0.51	51	0.51	51
Variable Costs	–	0.25	25	0.25	25	0.25	25
Capital Charges	–	–	–	1.08	108	2.70	270
TOTAL	–	12.76	1276	13.84	1384	15.46	1546

and variable operating costs of 25¢/bbl, these revenues are only adequate to cover processing costs, yielding a zero capital charge.

By contrast, if product prices average $16.62/bbl, product revenues of $1.546 million/day will be adequate to meet all processing costs and, in addition, return $270,000/day to capital recovery, or a 20% capital charge on the investment shown in Table 8-2.

Over an extended time period and in a free market environment, the prices of individual petroleum products should reflect the full investment required to construct new refineries, the full operating costs for these refineries, and the full cost of conversion of one product into another. The achievement of these price levels should be required to justify the new investments needed to meet future demand growth. For example, the price of gasoline should reflect some proportion of the total operating and capital cost of a new refinery, with that proportion being dictated primarily by the costs of conversion of fuel oil into gasoline. In the shorter term, however, supply/demand imbalances may create market forces that prevent the full recovery of refining costs. For example, if refinery capacity is generally in excess, it would not be expected that a new refinery would recover the full capital costs of its construction. Furthermore, government actions can interfere with these free market forces, preventing equilibrium prices of the products from being achieved. For example, price controls on gasoline may require reallocation of the refining costs to other refinery products, even when the full capital charge on the refinery is realized. Similarly, when there is a shortage of refining capacity, market prices may return higher levels of capital charges than would have been required to justify financing of the refinery.

Clearly, the econometric price models must be able to reflect not only future real changes in crude oil prices, but also future changes in overall refinery profitability, as well as individual processing unit profitability. This is achieved by constructing a series of equations (Table 8-3) for the four basic refinery types, relating processing costs to the composite prices of products produced. These equations, comprising the econometric model of the refining industry, are then solved simultaneously to provide product price projections.

In the first equation of Table 8-3, for example, the average price of petroleum products from any of the refinery types must be related to the total cost of processing, including crude costs and cash operating costs, as well as capital charges. If, for example, the available supply of conversion refineries greatly exceeds the demand for their products, or if government policies regarding product or crude pricing are unduly restrictive, little or no return on capital may be achieved. In this case, the zero percent capital charge entry of Table 8-2 may be applicable. In cases of even more extreme excess supply, not all operating costs may be recovered. Under these

Table 8-3. Equilibrium Price Relationships

Equation	Basis
1. Crude & Refining Costs = Value of All Products	All products jointly bear refining costs.
2. Naphtha & Reforming Costs = Gasoline	Gasoline-naphtha margin justifies reforming unit.
3. High-Sulfur Gas Oil + Desulfurization Costs = Low-Sulfur Gas Oil	Margin justifies desulfurization unit investment.
4. High-Sulfur Fuel Oil + Desulfurization Costs = 0.5% S Fuel Oil	Margin justifies desulfurization unit investment.
5. High-Sulfur Fuel Oil + Desulfurization Costs = 0.3% S Fuel Oil	Margin justifies desulfurization unit investment.
6. High-Sulfur Fuel Oil + Conversion Costs = Gas Oil & Gasoline	Margin justifies conversion unit investment.
7. Low-Sulfur Gas Oil + 0.5% S Fuel Oil = 0.3% Fuel Oil	Blending values relate gas oil to fuel oil.
8. Kerosene = Linear Interpolation of Naphtha and Gas Oil Values	Changing distillation cut points will adjust kerosene production.
9. Vacuum Bottoms + Gas Oil = High-Sulfur Fuel Oil	Blending values relate vacuum bottoms to fuel oil.

conditions, less efficient refineries would eventually be shut down, bringing supply more in balance with demand.

By contrast, a strong demand for this conversion refinery type is foreseen if, in a free market environment, margins will rise, allowing a sufficiently profitable operation to justify the construction of new refineries of this type, perhaps to the 25% capital charge level. In view of the rapid escalation of refinery construction costs (Figure 8-5), margins would have to increase substantially in future to provide the incentive for new refinery construction. As illustrated, the cost of building new refinery capacity of any type will be considerably higher than were the costs of installing currently operating refinery capacity.

PRODUCT PRICE FORECASTS
(CONSTANT 1978 DOLLARS)

The econometric refinery pricing model takes as input the refinery gate crude oil costs. The model contains data on the investment and operating costs of various refinery processing units, such as desulfurization and catalytic reforming. It also contains yield data on the crude oils, processing unit yield data and finished product blending equations. Table 8-4 shows the distillation yields for Arabian light crude oil. Table 8-5 presents the overall refinery yields established for each of the refinery types.

The model user provides several inputs about refining cost recovery and product price relationships. The forecast supply/demand balance for total

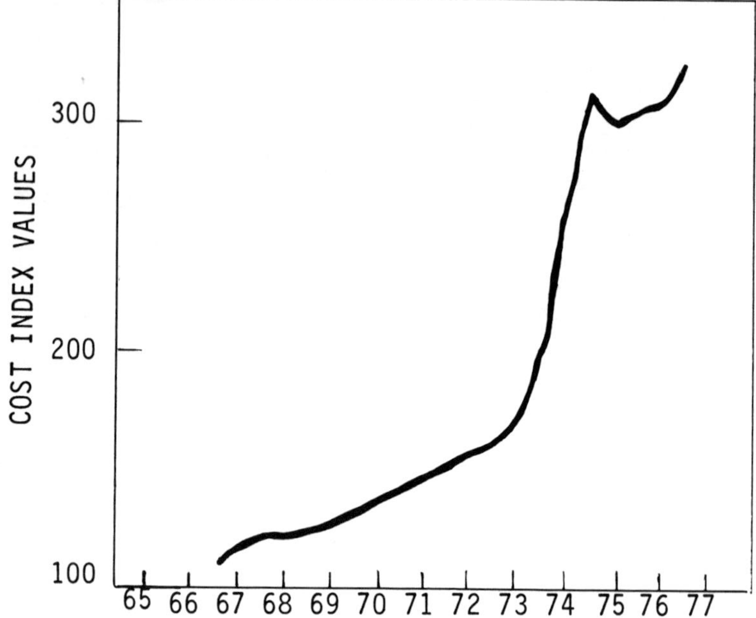

Figure 8-5. Refining and petrochemical process facilities total erected cost index (Sitter, 1977).

Table 8-4. Arabian Light Crude Assay

Product	Yield (LV %)
Ethane	—
Propane	0.17
Isobutane	0.17
Normal Butane	1.06
Straight Run Naphtha	
$\quad$ C$_5$ − 160°F	4.85
$\quad$ 160 − 375°F	22.15
Gas Oil 375 − 650°F	28.40
Fuel Oil 650°F +	43.20
Crude Oil Properties	
$\quad$ Gravity (°API)	34.5
$\quad$ % weight sulfur	1.70

refining capacity will have an impact on the level of cost recovery, including capital charges, which the refiner may obtain in a competitive market. the total refinery margin is the difference between the value of all products sold at the refinery gate (price times volume for each product) less the total cost

Table 8-5. Refinery Yields from Arabian Light Crude Oil (% of crude charge)

	Topping	Hydro-skimming	Hydroskimming and Desulfurization 0.5% S	0.3% S	Conversion with Desulfurization
Naphtha	27.0	–	(0.9)[a]	(1.3)	–
Gasoline	–	23.4	23.4	23.4	46.7
High-Sulfur Gas Oil	28.4	28.4	–	–	–
Low-Sulfur Gas Oil	–	–	28.1	28.1	29.8
High-Sulfur Fuel Oil	43.2	43.2	–	–	–
0.5% Sulfur Fuel Oil	(1.9)	(1.1)	41.2	(0.2)	16.5
0.3% Sulfur Fuel Oil	–	–	–	41.2	–
Refinery Fuel and Loss	1.4	5.0	7.3	7.3	7.0

[a]Figures in parentheses indicate purchase requirements for refinery fuel or for hydrogen production.

of crude oil delivered to the refinery. The program user is able to input his or her expectation of the level of total refinery cost recovery and, indirectly, of the level of refinery margin.

Given a fixed refined products production slate (number of barrels of gasoline, number of barrels of gas oil, etc.), any given level of total cost recovery may be obtained by an almost infinite number of combinations of prices for the individual products jointly produced from a single raw material—crude oil. The econometric model deals with this classical economic problem through the solution of a system of simultaneous equations, as described in Table 8-3. To set up the equations, the user specifies expectations about the levels of cost recovery, including capital charges, of individual units in the refinery. Specifically, these include the reforming fuel oil desulfurization and cracking units. As with the level of expected total refinery cost recovery, these inputs are based on forecasts of the relative supply and demand balances for the major projects. the reforming and conversion units are oriented toward gasoline production, while the desulfurization units convert high- to low-sulfur fuel oil. Thus, the demand outlook for these two products is especially important.

In the development of the cost allocation model, substantial effort was directed at calibration and validation of the model by comparing its price projections to historical data. However, projections of future prices are intended to represent long-term equilibrium trends of crude prices and refinery products supply and demand. These prices will not reflect short-term variations about major price trends due, for example, to seasonal fluctuations.

In this study, independent assessments of refinery supply/demand were not made; instead, recent projections by the Petroleum Industry Research Foundation, Inc. were used. From these studies, currently existing refinery capacity plus modest additions planned between now and the early 1980s will be sufficient to meet demand. Major expansions will be required by 1980. Gasoline demand growth rate is assumed to drop gradually from historic levels, due to fuel efficiency in automobiles. Kerosene and distillate demand growth is expected to be strong, reflecting air transportation, turbine power and petrochemical usage. Fuel oil growth is strong for five to eight years; by 1985, however, supplemental gas supplies will reduce this growth rate.

Investment requirements of the minor refinery debuggings and expansions assumed streamed between 1978 and 1985 are unlikely to exceed the costs of the major 1977 expansions and, therefore, will not exert an upward pressure on product prices, other than that due to increased crude oil prices. All of this capacity will tend to earn generally higher rates of return as demand and utilization increase concomitantly. The investments for 1990/1991 major expansion will exceed that of the 1977 expansions, even in constant 1977 dollars, as the rate of inflation in total refinery construction costs is assumed to exceed the general rate of inflation in the economy. Thus, by the early 1990s, product prices must rise to levels sufficiently attractive to justify the new higher cost investment in refining capacity. Throughout the 1990s this capacity will be the marginal source of refined products and will tend to earn higher returns as demand growth (and/or retirements of older capacity) permits higher rates of utilization.

From these assumptions, equilibrium product values, based on allocations of refining costs, are presented in Table 8-6. Note that these are refinery gate product values; however, transportation costs between the refinery and the petrochemical complex are negligible.

Table 8-6. Allocated Equilibrium Product Values as Dependent on Crude Price

Refinery Gate Arabian Light Crude Oil Price ($/bbl)	Refinery Gate Product Values ($/bbl)		
	Naphtha	Pyrolysis Gasoline	Pyrolysis Fuel Oil
12	15.0	17.8	12.8
14	17.2	20.0	15.6
16	19.2	22.1	18.0
18	21.3	24.3	20.0
20	23.3	26.5	22.1
22	25.2	28.7	23.9
24	27.2	30.8	26.0
26	29.2	33.0	27.8

CHAPTER 9

THE MITRE FULL LIFE-CYCLE COST MODEL

The MITRE Corporation, in conjunction with its work for electric utilities, has developed a computer model to project life-cycle operating costs for an electric generating plant and calculate a "cost" per unit of output. The model projects in considerable detail the actual cash expenditures for a privately owned corporation and includes the applicable income taxes. Since these cash flows are irregular, an equivalent uniform cash flow is constructed using the firms cost of capital as the common discount rate. This uniform cash flow is converted to a single cost per unit of product. Since the model evaluated electric generating technologies, the results were reported in mills/kWh.

A general expression for the unit cost determined by the model is:

$$P = \frac{\sum_{i=0}^{n-1} \sum_{k=1}^{m} C(k,i) \dfrac{[1 + E(k)]^i [1 + G]^i}{[1 + R]^{i+1}}}{\sum_{i=0}^{n-1} U(i) \dfrac{[1 + T]^i}{[1 + R]^{i+1}}} \tag{1}$$

where

$C(k,i)$	=	the annual cost of the kth cost component in the ith year, expressed in base year (i=0) dollars
$E(k)$	=	the annual price escalation rate for the kth cost component
G	=	the projected annual GNP deflator
R	=	the after income tax cost of capital to the firm
$U(i)$	=	the number of units of production in the ith year
T	=	the annual rate of increase in product price desired expressed as a fraction

> n   = the economic life of the plant in years
> m   = the number of cost components included
> p   = the "cost" per unit of output

The cash flows in Equation 1 are instantaneous year-end cash flows, and all rates of compounding are discrete. The equation may easily be changed to represent continuous cash flows and continuous compounding by substituting integral signs for the summations and the replacement of compound interest by the substitution of e^j for:

$$\lim_{s \to \infty} \left[1 + \frac{R}{s}\right]^s$$

The value of j now represents the force of interest. The MITRE model uses instantaneous year-end cash flow flows and discrete compounding.

The parameters in Equation 1 are more easily understood with specific examples. The C(k,i) cost components include all operating expenses: fuel, operating and maintenance costs, property taxes, etc. In addition, capital recovery and income taxes are included. Because of the complex U.S. tax laws, income taxes involve particularly tedious calculations. Unfortunately, eliminating this complexity results in a significant overstatement of the actual tax costs assigned to the product. The MITRE model includes these calculations, i.e., accelerated tax depreciation and the impact of the investment tax credit in the cost assessment.

The cost of capital in Equation 1 is a weighted average of the cost of debt and equity. The expression for the cost of capital is:

$$R = D \cdot I \cdot [1 - Z] + E \cdot S \tag{2}$$

where   D = average fraction of debt in the corporation
               I   = the interest rate for debt
               E = average fraction of equity in the corporation
               S = required after-tax rate of earnings for equity
               Z = the corporate income tax rate

Reasonable estimates for these parameters may be estimated by examining historical industry performance over the business cycle.

A uniform unit product cost projection over the lifetime of a plant during a period of high inflation presents a dilemma. If the product cost is constant, the inflation-adjusted cost of real cost declines over the plant life. As a result, the parameter T (see Equation 1) has been included in the model. If T equals 0, the product cost per unit in actual dollars remains constant and the real (inflation adjusted) cost declines. This is the typical assumption in unit cost estimating. However, if T equals the inflation rate, the unit cost increases each year at the rate of inflation, and the real cost is constant. In either case, the present value of the equivalent annual cost stream is equal to the present value of the actual costs. Therefore, under either method, the firm would just recover all costs, including return on investment, if the market price were to equal cost. When the unit dollar price is constant over the plant life, the firm recovers more than actual costs in early years and less than actual costs in later years. This is illustrated graphically in Figure 9-1.

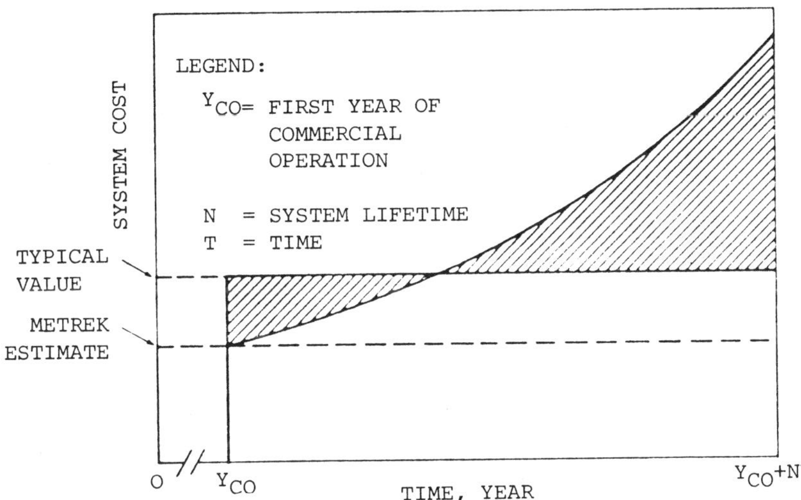

Figure 9-1. Comparison of approaches to leveled cost. Important items to note: (1) Both methods estimate the total revenue requirements in a comparable fashion and therefore result in the same estimate for total revenue requirements; (2) The MITRE method attempts to provide the index in a manner which closely resembles the anticipated growth in costs; (3) The more typical approach significantly overstates the cost in the early years and understates costs in latter years.

The MITRE Full Life-Cycle Costing Model generates two reports, which permit the user to verify the correctness of the cost parameters

REPORT 2   LIFE CYCLE COSTS   SEP 07, 1978   23:24

REGION : MID-CONTINENT U.S.   OLEFIN PLANT

INFLATION RATE 6%   TECHNOLOGY : NAPHTHA-CRACKING   UNIT COST :   34.032 CENTS

PROPERTY TAX : $7.88   INITIAL COST : $315.21   FEEDSTOCK : NAPHTHA - $15.00/BBL (78 PRICE)

YEAR	CASH FLOW	TAX DEPR.	DEBT INTER.	EARNINGS EQUITY	INCOME TAX	FINAN. DEPR.	FEED STOCK	O & M	PROP. TAX	TOTAL REV.	ETHY. OUTPUT
1979	46.83										
1980	98.34										
1981	103.26										
1982		70.05	9.55	39.55	-67.28	26.27	208.52	61.73	7.88	286.22	960
1983		61.29	8.76	36.25	1.23	26.27	221.03	65.43	7.88	366.85	960
1984		52.54	7.96	32.96	6.69	26.27	234.30	69.36	7.88	385.40	960
1985		43.78	7.16	29.66	12.15	26.27	248.35	73.52	7.88	404.99	960
1986		35.02	6.37	26.36	17.61	26.27	263.26	77.93	7.88	425.67	960
1987		26.27	5.57	23.07	23.07	26.27	279.05	82.60	7.88	447.51	960
1988		17.51	4.78	19.77	28.53	26.27	295.79	87.56	7.88	470.58	960
1989		8.76	3.98	16.48	33.99	26.27	313.54	92.81	7.88	494.95	960
1990		0.00	3.18	13.18	39.45	26.27	332.35	98.38	7.88	520.70	960
1991		0.00	2.39	9.89	36.15	26.27	352.29	104.29	7.88	539.16	960
1992		0.00	1.59	6.59	32.86	26.27	373.43	110.54	7.88	559.16	960
1993		0.00	0.80	3.30	29.56	26.27	395.84	117.18	7.88	580.82	960
P.V.	315.21	209.89	35.97	148.94	32.07	148.29	1513.35	447.98	44.48	2371.08	6967

NOTES: ALL COST AMOUNTS IN MILLIONS OF DOLLARS.   OUTPUT IN MILLIONS OF POUNDS.
NEGATIVE INCOME TAXES REDUCE TAXES ON INCOME EARNED BY OTHER CORPORATE INVESTMENTS.

Figure 9-2. Life cycle costs.

REPORT 3   LIFE CYCLE COST RECOVERY   SEP C7, 1978   23:24

REGION : MID-CONTINENT U.S.   OLEFIN PLANT

UNIT COST :   34.032 CENTS   FEEDSTOCK RATE   3.04878 POUNDS PER POUND ETHYLENE

PROPERTY TAX :   7.88   INITIAL COST :   315.21   FEEDSTOCK : NAPHTHA

YEAR	UNIT RATE	REVENUE	FEED STOCK	O & M	DEBT INT	TOTAL EXPS	BEFORE T. INC	INCOME TAXES	EQUITY EARN.	DEBT PAY	EQUITY PAY	DEBT BALANCE	EQUITY BALANCE
1981												95.50	219.70
1982	.34032	326.71	208.52	61.73	9.55	278.13	39.03	-47.02	39.55	14.09	32.42	81.42	187.28
1983	.36074	346.31	221.03	65.43	8.14	294.35	43.83	-8.72	33.71	5.71	13.14	75.71	174.15
1984	.38238	367.09	234.30	69.36	7.57	311.53	47.99	-2.26	31.35	5.73	13.18	69.97	160.96
1985	.40533	389.11	248.35	73.52	7.00	329.75	52.37	4.29	28.97	5.79	13.31	64.19	147.65
1986	.42964	412.46	263.26	77.93	6.42	349.06	56.98	10.98	26.58	5.89	13.54	58.30	134.11
1987	.45542	437.21	279.05	82.60	5.83	369.53	61.85	17.79	24.14	6.03	13.88	52.27	120.23
1988	.48275	463.44	295.79	87.56	5.23	391.23	66.98	24.74	21.64	6.24	14.36	46.02	105.87
1989	.51171	491.24	313.54	92.81	4.60	414.23	72.41	31.83	19.06	6.52	15.00	39.50	90.86
1990	.54242	520.72	332.35	98.38	3.95	438.62	78.16	39.08	16.36	6.89	15.84.	32.61	75.03
1991	.57496	551.96	352.29	104.29	3.26	464.46	84.24	42.12	13.50	8.67	19.95	23.94	55.08
1992	.60946	585.08	373.43	110.54	2.39	491.85	90.83	45.42	9.91	10.76	24.75	13.19	30.33
1993	.64603	620.18	395.84	117.18	1.32	520.89	97.98	48.99	5.46	13.19	30.33		

| | FEEDSTOCK | O & M | TAXES | | | CAPITAL |
			INCOME	OTHER	TOTAL	RECOVERY
PV. DOLLARS	1,513.35	447.98	31.83	44.49	76.32	333.44
CENTS/POUND	21.721	6.430	.457	.639	1.095	4.786
PERCENT	63.83	18.89	1.34	1.88	3.22	14.06

Figure 9-3. Life cycle cost recovery.

```
REPORT 2                    LIFE CYCLE COSTS                    SEP 07, 1978   23:24

REGION : MID-CONTINENT U.S.          OLEFIN PLANT

INFLATION RATE 6%        TECHNOLOGY : NAPHTHA-CRACKING      UNIT COST :   43.752 CENTS

PROPERTY TAX : $7.88    INITIAL COST :   $315.21    FEEDSTOCK : NAPHTHA - $15.00/BBL (78 PRICE)
```

YEAR	CASH FLOW	TAX DEPR.	DEBT INTER.	EARNINGS EQUITY	INCOME TAX	FINAN. DEPR.	FEED STOCK	O & M	PROP. TAX	TOTAL REV.	ETHY. OUTPUT
1979	46.83										
1980	98.34										
1981	103.26										
1982		70.05	9.55	39.55	-67.28	26.27	208.52	61.73	7.88	286.22	960
1983		61.29	8.76	36.25	1.23	26.27	221.03	65.43	7.88	366.85	960
1984		52.54	7.96	32.96	6.69	26.27	234.30	69.36	7.88	385.40	960
1985		43.78	7.16	29.66	12.15	26.27	248.35	73.52	7.88	404.99	960
1986		35.02	6.37	26.36	17.61	26.27	263.26	77.93	7.88	425.67	960
1987		26.27	5.57	23.07	23.07	26.27	279.05	82.60	7.88	447.51	960
1988		17.51	4.78	19.77	28.53	26.27	295.79	87.56	7.88	470.58	960
1989		8.76	3.98	16.48	33.99	26.27	313.54	92.81	7.88	494.95	960
1990		0.00	3.18	13.18	39.45	26.27	332.35	98.38	7.88	520.70	960
1991		0.00	2.39	9.89	36.15	26.27	352.29	104.29	7.88	539.16	960
1992		0.00	1.59	6.59	32.86	26.27	373.43	110.54	7.88	559.16	960
1993		0.00	0.80	3.30	29.56	26.27	395.84	117.18	7.88	580.82	960
P.V.	315.21	209.89	35.97	148.94	32.07	148.29	1513.35	447.98	44.48	2371.08	5419

```
NOTES: ALL COST AMOUNTS IN MILLIONS OF DOLLARS.  OUTPUT IN MILLIONS OF POUNDS.
       NEGATIVE INCOME TAXES REDUCE TAXES ON INCOME EARNED BY OTHER CORPORATE INVESTMENTS.
```

Figure 9-4. Life cycle costs.

```
REPORT 3                        LIFE CYCLE COST RECOVERY              SEP 07, 1978  23:24

REGION : MID-CONTINENT U.S.     OLEFIN PLANT

UNIT COST :  43.752 CENTS       FEEDSTOCK RATE  3.04878 POUNDS PER POUND ETHYLENE

PROPERTY TAX :     7.88         INITIAL COST :    315.21        FEEDSTOCK : NAPHTHA
```

YEAR	UNIT RATE	REVENUE	FEED STOCK	O & M	DEBT INT	TOTAL EXPS	BEFORE T. INC	INCOME TAXES	EQUITY EARN.	DEBT PAY	EQUITY PAY	DEBT BALANCE	EQUITY BALANCE
1981												95.50	219.70
1982	.43752	420.01	208.52	61.73	9.55	278.13	132.34	-.37	39.55	28.23	64.94	67.28	154.77
1983	.43752	420.01	221.03	65.43	6.73	294.35	118.94	28.83	27.86	18.86	43.40	48.42	111.37
1984	.43752	420.01	234.30	69.36	4.84	311.53	103.64	25.55	20.05	17.59	40.46	30.83	70.92
1985	.43752	420.01	248.35	73.52	3.08	329.75	87.18	21.70	12.76	15.97	36.74	14.86	34.17
1986	.43752	420.01	263.26	77.93	1.49	349.06	69.47	17.22	6.15	13.97	32.13	.89	2.05
1987	.43752	420.01	279.05	82.60	.09	369.53	50.39	12.06	.37	11.50	26.46	-10.60	-24.40
1988	.43752	420.01	295.79	87.56	-1.05	391.23	29.84	6.17	-4.38	8.51	19.57	-19.11	-43.97
1989	.43752	420.01	313.54	92.81	-1.90	414.23	7.69	-.52	-7.91	4.89	11.25	-24.00	-55.22
1990	.43752	420.01	332.35	98.38	-2.39	438.62	-16.19	-8.09	-9.93	.56	1.28	-24.56	-56.51
1991	.43752	420.01	352.29	104.29	-2.45	464.46	-41.98	-20.98	-10.16	-3.27	-7.53	-21.28	-48.96
1992	.43752	420.01	373.43	110.54	-2.12	491.85	-69.70	-34.84	-8.81	-7.88	-18.14	-13.39	-30.82
1993	.43752	420.01	395.84	117.18	-1.33	520.89	-99.53	-49.76	-5.54	-13.39	-30.82		

	FEEDSTOCK	O & M	INCOME	TAXES OTHER	TOTAL	CAPITAL RECOVERY
PV. DOLLARS	1,513.35	447.98	41.96	44.49	86.45	323.30
CENTS/POUND	27.925	8.266	.774	.821	1.595	5.966
PERCENT	63.83	18.89	1.77	1.88	3.65	13.64

Figure 9-5. Life cycle cost recovery.

entered to project plant costs and also to illustrate the precise recovery of all costs over the life of the plant. These reports are presented for the Naphtha Cracking Process in Figures 9-2 through 9-5 under the assumptions of a unit price increasing with inflation in Figures 9-2 and 9-3 and a constant unit price in Figures 9-4 and 9-5. The model headings have been corrected to reflect the chemical conversion process instead of an electric plant.

CHAPTER 10

HISTORIC COST/PRICE TRENDS

To examine price trends of compounds associated with the production of ethylene and VCM, a price history for each was prepared. The data cover 1970–1977. Data for 1978 are presented for completeness. Data for a total of 20 compounds were collected and are presented on a $/lb basis (Table 10-1). For some of the compounds there are gaps in the data for the years 1973, 1974, 1975, 1977 and 1978. No information is present for ethane, a component of natural gas, since most of the natural gas is sold interstate, where the price is regulated.

For ease of presentation, the data are broken into three periods: 1970–1972, 1973–1974 and 1974–1977. No attempt is made here to explain the price behavior for the years considered, except to point out the dramatic price increase after 1973. Table 10-1 shows that after the oil embargo and foreign crude oil price increase in 1973, prices for all the compounds began to rise, including those associated with natural gas or obtained from other sources. The fourfold price increase in 1973–1974, was cushioned by existing oil reserves. The full impact was felt in 1974–1975. An important factor in restraining price increases for oil and its derivatives is that U.S. crude is not permitted to rise in price as fast as foreign crude because of the regulations controlling the sale of petroleum in the U.S.

From 1972–1974 the prices of compounds jumped more than 100%. Subsequent years have seen price increases, but not as dramatic as the ones for 1972–1974. In that same period, prices for aromatic compounds increased an average of 227%; toluene rose 256% from $0.023 in 1972 to 0.082 in 1973. The price of benzene increased from 0.028 to 0.090, a 221% price increase. Other aromatic compounds went up 44%, from 0.018 to 0.026.

Table 10-1. Price History of Selected Compounds Used in the Petrochemical Industry
(prices of olefin plant products, $/lb)

Compounds	Year								
	1970	1971	1972	1973	1974	1975	1976	1977	1978[a]
Benzene	0.030	0.027	0.028	0.040	0.090	0.096	0.106	0.116	0.103
Toluene	0.025	0.023	0.023	0.032	0.082	0.061	0.076	0.085	0.082
Ethylene	0.031	0.030	0.030	0.033	0.075	0.088	0.112	0.112	0.125
Propylene	0.027	0.027	0.029	0.028	0.069	0.070	0.074	0.100	0.100
Butadiene	0.084	0.083	0.078	0.081	0.145	0.165	0.176	0.190	0.210
Vinyl Chloride Monomer	0.048	0.048	0.048	0.050	N/A[b]	N/A	0.131	0.143	0.141
Hydrogen Chloride	0.06	0.06	0.06	0.06	N/A	N/A	0.085	0.085	0.090
Chlorine	0.038	0.038	0.038	0.038	N/A	N/A	0.068	0.068	0.068
Acetylene	0.091	0.113	0.112	0.130	0.175	0.260	0.257	N/A	N/A
Other Aromatics	0.018	0.018	0.018	0.019	0.026	0.051	0.059	N/A	N/A
C_3	0.013	0.013	0.014	0.018	0.043	0.055	0.073	N/A	N/A
C_4	0.028	0.026	0.027	0.026	0.064	0.072	0.080	N/A	N/A
Coal	N/A	N/A	N/A	N/A	N/A	0.012	0.010	0.012	N/A
Crude Oil[c]	N/A	N/A	N/A	N/A	0.027	0.031	0.033	0.036	0.037

[a]Partial data.
[b]Not available.
[c]Partial data on 1975 and 1978.

Alkenes increased in price as well, the average being 107%, with ethylene taking the biggest share. The price for ethylene increased 150% from $0.030 in 1972 to $0.075 in 1974. In 1972, the price of propylene went up 137% to 0.069 from 0.029, and the price of butadiene jumped to 0.146 from 0.078, an 87% increase. Acetylene posted a 56% increase—to 0.175 in 1974 from 0.112 in 1972.

Data for VCM are not available for this period; however, its price went up in the same fashion as the other compounds. From 1973 to 1976, the price of VCM jumped 162% to 0.131 from 0.050. Small-chain hydrocarbons (C_3 and C_4) also registered price increases for this period; the price of C_3 went up from 0.014 to 0.043, a 207% increase. C_4 posted increases that were not as spectacular as those for C_3; they jumped 137% to 0.064 from 0.027. Data for inorganic compounds (hydrogen chloride and chlorine) show gaps in 1974 and 1975. From 1973 to 1976, the price of hydrogen chloride (HCl) increased 42% to 0.085 from 0.060; for the same period, chlorine (Cl_2) went up 79% from 0.038.

After 1974, prices began to stabilize for some compounds, whereas for others, the variations remained pronounced. From 1974 to 1977, benzene went from 0.09 to 0.116. The increase came in 9% yearly increments. Toluene had significant price variations in this period; after a price dive to 0.061 in 1975 from 0.082 in 1974, it went up to 0.076 in 1976 and finally posted an 11% increase to 0.085 in 1977. Other aromatics went up 96% from 0.026 to 0.051 in 1975 and registered a 15% increase in 1976. Prices for ethylene increased 17% to 0.088 from 0.075; subsequently, they jumped 24% to 0.112 in 1976, where the price stabilized. The price for propylene for this period posted a gain of 1% to 0.070 from 0.069 in 1974; in 1976, it increased to 0.074, and finally to 0.100 in 1977. The butadiene price increased 13% to 0.165 in 1975 and jumped to 0.176 in 1976, an increase of 7%. In 1977, the price for butadiene increased 8%. VCM posted a 9% price increase from 0.131 to 0.143 in 1977. The price for acetylene jumped 48% to 0.260 in 1975 from 0.175 in 1974. It had a modest price decrease in 1976 to 0.257.

C_3 went up 27% in 1974 from 0.043 to 0.055 and 32% to 0.073 in 1977. C_4 went up 12% in 1974 from 0.064 to 0.072 and 11% to 0.080 in 1977. The inorganic compounds and coal remained stable during this period. Crude oil's price increased 33% to 0.036 in 1977 from 0.027 in 1974. The increases were at the average annual rate of 8%. The rate of price increase during this period was not as great as during the 1972–1974 period.

REFERENCES

Barbier, M., J. Vlahakis, R. Ouellette, R. Pikul and R. Rice. "Radiation Curing," in *Electrotechnology, Vol. 2. Applications in Manufacturing*, R. P. Ouellette, F. Ellerbusch and P. N. Cheremisinoff, Eds. (Ann Arbor, MI: Ann Arbor Science Publishers, Inc., 1978), pp. 29–62.

"Can Carbide's Reactor Cut the Crude?" *Chem. Wk.* (May 11, 1977).

"Carbide Details Ethylene From Crude Process," *Chem. Eng. News* (March 27, 1977).

"Crude Crackers Are Not For Everyone," *Chem. Wk.* (February 1, 1978).

"DOW, UCC Crude Cracking Overtakes European Work," *Eur. Chem. News* (December, 1977).

DRAVO. "1978 Economic Evaluation of the AVCO Arc-Coal Acetylene Process," Chemical Plants Division, Pittsburgh, PA (March 1978).

"Ethylene: The End of an Era," *Chemical Engineering* (March 28, 1977).

"European Ethylene: Slow Growth Ahead," *Chemical Week* (March 20, 1977).

Gomi, S., and L. A. Wilkinson. "Cracking Crude for Olefins Cuts Costs," *Oil Gas J.* 59-61 (June 1974).

Goudarzi, L., R. Gilbert, R. Kuehnel and B. Buswell. "Analysis of Benefits Associated with the Introduction of Advanced Generating Technologies," The MITRE Corporation, MTR 7388 (March 1977).

Kienkler, R. W. "Overview," The Executive Briefing on Feedstocks for Chemicals, The Energy Bureau (1978).

Guccione, E. "Coal's Improving Outlook as a Source of Chemicals," *Coal Mining Proc.* 58-84 (August 1977).

Medville, D., J. Roseber, D. Salo and M. Shauffler. "Comparative Economic Assessment of Ethanol from Biomass," The MITRE Corporation, MTR-7936 (1978).

Minet, R. G., and F. W. Tsai. "Feedstock Outlook Changing in U.S.," *Oil Gas J.* 135-141 (March 7, 1977).

"New Doubts About Coal-Based Chemicals," *Chem. Eng. News* (November 21, 1977).

"New Trends in Olefin Feedstocks," *Oil Gas J.* (October 18, 1971).

Park, W., G. Paice and D. Salo. "Biomass-Based Alcohol Fuels," The MITRE Corporation, MTR-7866 (1978).

Schaffer, F. P., & Associates, Inc. "Economic Study of Alcohol from Cane Juice," Baton Rouge, LA.

Sitter, C. R. "The Automotive News World Congress," Detroit, Michigan (July 11, 1977).

Struth, B. W. "Economics and Pricing," The Executive Briefing on Feedstocks for Chemicals, The Energy Bureay (1978).

Struth, B. W. "Price and Availability," The Executive Briefing on Feedstocks for Chemicals, The Energy Bureau (1978).

Tsao, U., and J. W. Reilly. "A Fluid Bed Process for Ethylene from Ethanol," paper presented at the 70th Annual Meeting, American Institute of Chemical Engineers, New York, 1977.

Yang, V., and N. R. Juchi. "AGRO—Industrial System from Ethanol and Ethylene Production," *World Conference and Exhibition on Energy from Biomass and Waste*, Washington, DC (1978).

Yu, G., M. Mamadalier, M. M. Mamedov, M. R. Gusseinov, Sharifovia, and F. A. Mekhtieva. *Dokl. Akad. Nauk SSSR* 144(6):1303 (1962).

BIBLIOGRAPHY

"AVCO Arc-Coal Process," AVCO Corporation (PB-234 386) (August 1968).

Baba, T. B., and J. R. Kennedy. "Ethylene and Its Coproducts: The New Economics," *Chem. Eng.* (January 5, 1976).

Blaw-Know Chemical Plants, Inc. *Review and Evaluation of 300MM Lbs/Yr Acetylene Plant AVCO Arc-coal Process*, Research and Development Report No. 67, prepared for the Office of Coal Research, U.S. Department of the Interior, Washington, DC (1971).

Bus. Wk. 84-88 (July 10, 1978).

Cahill, A. M. *Chem. Eng. Prog.* 26-28 (July 1977).

National Energy Board. "Canadian Oil, Supply & Requirements," (February 1977).

"Canadian Renewable Fuels Study," *Energy Resources Technol.* (August 25, 1978).

"Challenges and Uncertainties Face Growing Brazilian Alcohol Fuels Program," *Energy Resources Technol.* (August 25, 1978).

Chemical Marketing Reporter, *Chem. Profile*, 1970-1978.

Colding, B. S. "Olefin Pricing in Europe," *Chem. Eng. Prog.* (July 1977).

"Deriving Chemicals from Trees Examined," *Chem. Eng. News* (July 25, 1977).

Dosher, J. R. "Changing Petrochemical Feedstocks—Causes and Effects," *Chem. Eng. Prog.* (September 1976).

"Dramatic Cost Reductions for Biomass/Alcohol Fuels Seen in Membrane Technology," *Energy Resources Technol.* (November 1978).

"Environmental Considerations of Selected Energy Conserving, Manufacturing Process Options," Volume VI, Olefins, EPA-600/7-76-034F.

"Ethylene: The End of an Era," *Chem. Eng.* (March 28, 1977).

"European Ethylene: Slow Growth Ahead," *Chem. Wk.* (March 26, 1977).

Farah, O. G., M. Meader, G. Mouchahoir and D. Nainan. "Overview of Western Hemisphere Energy, Basic Data and Statistics," The MITRE Corporation, M78-87, Rev. 1 (1978).

"Gulf Chemical's Biomass to Ethylene Plant cost Competitive with Petrochemicals," *Energy Resources Technol.* (July 21, 1978).

Hatch, L. F., and S. Matar. "From Hydrocarbons to Petrochemicals," *Hydrocarbon Proc.* (May 1977).

"Industrial Process Profiles for Environmental Use: Basic Petrochemicals Industry," No. 5, EPA-600/2-77-023C.

James, D., and M. Mendis. "Impact Analysis of the Arc-Coal Process Acetylene Technology," DOE Contract EC-77-C-03-1701, Hittman Associates (1978).

Kienker, R. W. "Overview—Executive Briefing on Feedstocks from Chemicals," The Energy Bureau (1978).

Kirk-Othmer. *Encyclopedia of Chemical Technology* (New York: John Wiley & Sons, Inc., 1964).

Kirkpatrick, D. M. "Acetylene from Calcium Carbide, an Alternate Feedstock Route," *Oil Gas J.* (June 7, 1976).

Miccolis, J. M. F. "Alternative Energy Technologies in Brazil," *Intersiencia* 3(5) (September-October 1978).

"The Oil Companies are Sloshing in Ethylene," *Bus. Wk.* (July 10, 1978).

"Oil Farming is Attracting Attention of Big Business," *Inside R&D* 7(45) (November 8, 1978).

Prescott, J. H. *Chem. Eng.* 63-65 (March 28, 1977).

Rao, K. V., and I. Skeist. "Coal Derived Chemicals Could Open New Frontiers," *Oil Gas J.* 90-93 (February 12, 1971).

Rothman, S. N., and D. E. Crouse. "Incentives Could Change Foreign Ethylene Picture," *Oil Gas J.* (March 21, 1977).

Struth, B. "Economics and Prices, Executive Briefing on Feedstocks from Chemicals," The Energy Bureau (1978).

Tucker, W., and M. A. Abrahams. "Economics of Petrochemical Output Changes," *Oil Gas J.* 81-84 (April 11, 1977).

"U.S. Ethylene to grow at 6.5%/Year," *Oil Gas J.* (November 1976).

Union Carbide Corporation. *Union Carbide Digest*, No. 77-10 (April 1977).

Woodhouse, G. J. "Ethylene Feedstocks to Change in Europe as Supply Fluctuates," *Oil Gas J.* (August 2, 1976).

INDEX

acetylene 27,28,37,39,46,50
  from coal (arc process) 43-45,55,56,
    63,72,73
  to vinyl chloride monomer 45-47
acrylic acid 27
acrylonitrile 27
Advanced Cracking Reactor (ACR) 3,
    6,8,71
alternate feedstocks, independent vari-
    ables 38,39
Arabian light crude 7,75,83
arc-cool plant 46
arc-cool process 28,49

BASF process 28
biomass 1,14-16,18,20,23,37,38,41,72
BTX 20
butadiene 13,20

capacity 1,34
capital costs 50
capital investment 49
catalytic cracking 6
catalytic dehydration 26

cellulose 16
C. E. Lummus Co. 72
chemical process cost flow 50
China 18
Chiyoda Chemical Engineering & Con-
    struction Co. 71
Chiyoda Kako Kensetsu 3
coal 14,15,18,23,55
  conversion 44
  -derived feedstock 20
  reactor 44
coke production 5
computer model 58
computer simulations 11
construction 3
consumption of ethylene 25
conventional processes, improvements
    5,6
conversion refinery 78,79
cost allocation 49,75-81
cost
  of acetylene 50,61
  of crude oil 51
  of ethylene 60,66,68
  of feedstock 51
  of vinyl chloride 65,66

cost/price trends 93-95
cracked gases 9
crude oil 37
crude price 54

desulfurization 77
diolefins 5
direct cracking 6-12
Dow Chemical 11,14

econometric price models 80
econometric refinery pricing 81
economic evaluation 37,45
electricity 39
electrolysis of salt 47
end uses of ethylene 26
ethanol 25,26
   plant 43,54
   production 41
ethyl benzene 25
ethylene
   capacity 30
   dichloride 25,47
   from biomass 72
   from coal 72,73
   from crude oil 38,40,71,72
   from naphtha 51-53,62
   from sugar cane 41-43,53-55
   oxide 25
   production 28
   to vinyl chloride monomer 47,48
European ethylene capacity 35,36

feedstocks 1,28,29,34
   for ethylene 31
   sources 30
fermentation 41,43
financial parameters 52
fuels refinery schematic 77

gas liquids supply 29
gas oil feedstocks 32
gasoline production 76
geopressured reservoirs 3
glucose 16
growth rates 13

heavy oils 2,22
high-sulfur fuel oils 77
hydrogenated oil 18
hydroskimming 77

income tax 51
investments 18
Iran 18

Kureha Chemical Co. 3,71

liquefied natural gas 16

market conditions 2
Mexico 18
MITRE cost model 85
molasses 43

natural gas 3,16-19
   prices 20
naphtha 21,32,40,75
   cracker 10,20
   feedstock 40,75
new processes 6

oil prices 18,39
oil shortages 21;
olefin 5
   plant yields 32,33,40
OPEC 1,14
Orbach reactor 73
Ozaki quench cooler 7

petrochemical feedstock 15
planning 3
polyethylene 25
price
   allocation formulas 31
   of electricity 39
   of natural gas 22
   of oil 19
   relationships 81
procedure for cost comparisons 49-51
product costs 59

product price forecasts 81-84
product values 54,85
production capacity 17
projections 25
propylene 13,20
pyrolysis 5

quenched reactor products 7

raw materials 14-23
refinery cost allocation 75
refinery yields 83
revenue requirements 49

sources of supply 13

tar sands 2,22

technology improvement 71-73
trends 28-35

Union Carbide 3,11,14
unit cost 85

variable costs 52
VCM technology 69
vinyl acetate 27
vinyl chloride 27,37,56,57,93
   from acetylene 56,57
   from ethylene 56,57

world ethylene industry 35,36
Wulff process 28

yields 10,31-35